Project Quality Management

Sui Pheng Low · Joy Ong

Project Quality Management

Critical Success Factors for Buildings

Springer

Sui Pheng Low
Joy Ong
Department of Building
National University of Singapore
Singapore
Singapore

ISBN 978-981-10-1245-7 ISBN 978-981-287-074-2 (eBook)
DOI 10.1007/978-981-287-074-2
Springer Singapore Heidelberg New York Dordrecht London

Softcover reprint of the hardcover 1st edition 2014

Printed on acid-free paper

Springer is part of Springer Science+Business Media (www.springer.com)

Contents

1 Introduction 1
1.1 Background 1
1.1.1 Quality Management System 1
1.1.2 International Organisation for Standardisation 9000 2
1.1.3 Construction Quality Assessment System 3
1.2 Research Problem and Hypothesis 3
1.3 Research Aim and Objectives 5
1.4 Significance of Research 8
1.5 Book Structure 8
References 8

2 Construction Quality Assessment System 11
2.1 Evolution of CONQUAS 11
2.1.1 History and Development 11
2.1.2 Bonus Scheme for Construction Quality 12
2.1.3 Introduction of Quality Mark Scheme 14
2.2 Assessment Approach 16
2.3 Sampling Approach 17
2.4 Quality Trends 18
2.4.1 Mechanical and Electrical Works 19
2.4.2 Architectural Works 20
2.4.3 Structural Works 23
2.5 Summary of CONQUAS Findings 25
References 26

3 Quality Management 29
3.1 Evolution of Quality Management 29
3.2 Rationale for Quality Management in the Construction Industry 30
3.3 Quality Control and Quality Assurance 31
3.4 Pareto Efficiency of Quality Management 31
3.5 Barriers to Quality Management in Construction 32
3.6 Application of Quality Management to CONQUAS 33
References 35

4 CONQUAS Critical Success Factors 37
4.1 Introduction 37
4.2 Human Resource Management 38
4.2.1 Formation of Quality Management Unit with Clear Guidelines of Each Member's Roles and Responsibilities for Every Project: HRM1 38
4.2.2 Commitment of Top Management to Advance the CONQUAS Management Performance: HRM2 40
4.2.3 Getting the Support and Cooperation of Employees: HRM3 41
4.2.4 Training Employees to Understand and Adhere to CONQUAS Requirements: HRM4 42
4.2.5 Supervision by Client or Architect to Monitor the CONQUAS Management Workflow of the Main Contractor: HRM5 43
4.2.6 Comparing Company-Wide CONQUAS Performance Track Records: HRM6 44
4.2.7 Setting a Benchmark to Gauge and Control the CONQUAS Performance: HRM7 44
4.3 Subcontract Management 46
4.3.1 Controlling the Number of Subcontractors Working on the Project: SCM1 46
4.3.2 Having a Rigorous Prequalification Process to Select Subcontractors and Suppliers: SCM2 47
4.3.3 Ensuring the Skill Level of Labourers: SCM3 47
4.3.4 Collaborative Efforts Between Subcontractors and Main Contractor: SCM4 48
4.3.5 Adequacy of Contract Period and Contract Sum: SCM5 48
4.3.6 Identifying a Specific CONQUAS Score that the Subcontractor is Liable for Achieving: SCM6 49
4.3.7 Awarding Incentives to Subcontractors if Their Targeted Score is Met: SCM7 49
4.3.8 Imposing a Penalty to Subcontractors if Their Targeted Score is not Achieved: SCM8 49
4.3.9 Giving Clear Instructions to Subcontractors on How to Adhere to the CONQUAS Requirements: SCM9 49
4.4 Schedule Management 50
4.4.1 Prefixing of CONQUAS Sample Locations (Applicable to Structural Components Only): SM1 50
4.4.2 Adequacy of Sample Locations Prepared in the Event of Unforeseen Site Conditions to Serve as a Backup: SM2 51
4.4.3 Arranging for Phased Construction to Reap Learning Curve on Workmanship Quality: SM3 52

4.4.4 Booking of Assessment Schedule to Tie in with Site Progress: SM4 52
4.5 Material Management. 52
4.5.1 Choose Materials Through a Comparison Process, Considering Their Specifications and Samples to be Provided as Well: MM1 53
4.5.2 Select Materials Which can be Better Managed During Construction: MM2 54
4.5.3 Inspect Materials Upon Delivery: MM3 54
4.5.4 Proper Materials Handling and Storage: MM4 55
4.5.5 Protection of Materials After Completion of that Portion of Works: MM5 55
4.5.6 Sample Testing of Materials Through an Independent Testing Agency (ITA) to Check for Proper Usage of Materials: MM6 55
4.6 Construction Management 56
4.6.1 Ensuring that Shop Drawings are Checked Thoroughly Before Actual Construction: CM1 56
4.6.2 Constructing Mock-Ups to Check for Implications with Other Trades of Works: CM2 57
4.6.3 Field Demonstration by Labourers to Showcase Their Understanding of the Workmanship Quality Required: CM3 58
4.6.4 Conducting Preparatory Inspection Using Template Checklist at Every Stage of Work Activity: CM4 58
4.6.5 Using Recognised Testing Standards to Check for Non-compliance During Inspection: CM5 61
4.6.6 Cross-Checking by Another Party: CM6 61
4.6.7 Adherence to Reporting and Follow-Up Procedure of Defects Before CONQUAS Assessment: CM7 62
4.7 Summary of Critical Success Factors 62
References 65

5 Research Design and Methodology 69
5.1 Overview of Research Process 69
5.2 Survey Design 70
5.2.1 Type of Respondents 70
5.2.2 Survey Questions 73
5.2.3 Methodology for Survey Analysis 74
5.3 Interview Design 75
5.3.1 Interviewee's Profile 75
5.3.2 Interview Questions 75
5.4 Case Study Design 76
References 77

6 Results and Analysis . . . 79
6.1 Overview of Research Analysis . . . 79
6.2 Survey Response Rate and Demographics . . . 79
6.3 Importance Level of the Critical Success Factors . . . 79
6.3.1 Mean Importance Rating Based on the Five Categories . . . 85
6.4 Usage Level of the Critical Success Factors . . . 87
6.4.1 Mean Usage Rating Based on the Five Categories . . . 89
6.5 Correlation Analysis . . . 89
6.5.1 Coefficient of Determination . . . 89
6.5.2 Pearson Correlation . . . 92
6.5.3 Concluding Statement . . . 96
6.6 Linear Regression . . . 96
6.7 Independent-Samples Test . . . 100
6.7.1 Importance Rating . . . 101
6.7.2 Usage Rating . . . 107
6.8 Summary of Analysis . . . 109
References . . . 114

7 Case Study . . . 115
7.1 Outline of Company Z . . . 115
7.2 CONQUAS Background . . . 116
7.3 Early Period (Before Year 2002) . . . 117
7.4 Transition Period (Year 2002–2008) . . . 117
7.5 Current Practices (Year 2008 Onwards) . . . 118
7.6 Outcome . . . 119
7.7 Lessons Learnt . . . 120
Reference . . . 120

8 CONQUAS Management Framework . . . 121
8.1 Model for CONQUAS Management . . . 121
8.2 Stage A: Input Planning . . . 122
8.2.1 Stage A1: Human Resource Management . . . 122
8.2.2 Stage A2: Subcontracts Management . . . 123
8.2.3 Stage A3: Schedule Management . . . 123
8.3 Stage B: Construction Planning . . . 126
8.4 Output of the Framework . . . 130
8.5 Application . . . 132
8.6 Validation of the Framework . . . 132
References . . . 133

9 Conclusion . . . 135
9.1 Discussion of Main Findings . . . 135
9.1.1 Objective 1: To Examine the Development of CONQUAS . . . 135

9.1.2 Objective 2: To Investigate the Application of QMS to Manage CONQUAS . . . 136
9.1.3 Objective 3: To Identify CSFs for Achieving High CONQUAS Scores . . . 137
9.2 Validation of Hypotheses . . . 138
9.3 Implications to the Industry . . . 139
9.4 New CONQUAS 2012 Requirements . . . 140
9.5 Limitations and Challenges of Study . . . 141
9.6 Recommendations for Future Research . . . 142
9.7 Concluding Remarks . . . 143
References . . . 143

Appendix A: Survey Questionnaire . . . 145

Appendix B: Verbatim Report 1 . . . 149

Appendix C: Verbatim Report 2 . . . 153

Appendix D: Verbatim Report 3 . . . 157

Appendix E: Statistically Significant Positive Correlation for Importance Rating . . . 161

Appendix F: Statistically Significant Positive Correlation for Usage Rating . . . 167

Appendix G-1: Laying of Brickwall Inspection Template . . . 175

Appendix G-2: Prestressed Concrete Structure Inspection Template . . . 177

Appendix G-3: Cast In-Situ Concrete Inspection Template . . . 179

Appendix G-4: Precast Concrete Structure Inspection Template . . . 181

Appendix G-5: Structural Steelwork Inspection Template . . . 183

Bibliography . . . 185

Abbreviations

BCA	Building and Construction Authority
BSCQ	Bonus Scheme for Construction Quality
BSI	British Standard Institution
CEA	Construction Excellence Awards
CM	Construction Management
CONQUAS	Construction Quality Assessment System
CSF	Critical Success Factor
eNPQS	Electronic National Productivity and Quality Specifications
GIP	Good Industry Practices
HRM	Human Resource Management
IQUAS	Information on Construction Quality
ISO	International Organisation for Standardisation
ITA	Independent Testing Agency
M&E	Mechanical and Electrical
MM	Materials Management
PDCA	Plan Do Check Act
QA	Quality Assurance
QC	Quality Control
QCC	Quality Control Circle
QM	Quality Mark
QMS	Quality Management System
QMU	Quality Management Unit
SCM	Subcontracts Management
SM	Schedule Management
TQM	Total Quality Management
WTT	Water Tightness Test

Abstract

Intensifying global competition and increasing demand by clients for better quality have caused more and more companies to realise that they will have to provide quality products and/or services in order to successfully compete in the marketplace. However, good quality is hard to achieve and sustain. Ensuring workmanship quality is tough, particularly in the construction sector, where clients expect the final built product to be of high quality but low cost and constructed in the shortest time possible. Hence, contractors are facing increasing complexities to improve workmanship performance.

According to Deming, without measuring something, it is impossible to improve it and this means that one needs to determine the quality management criteria and measure its effect on workmanship performance. In Singapore, there is a defined quality performance measurement system called the Construction Quality Assessment System (CONQUAS) which has proven to be a challenging task for contractors to achieve a high score. There is therefore, a practical value, of researching on the Critical Success Factors (CSFs) for achieving high CONQUAS scores by contractors.

This book primarily focuses on quality issues involved and the 33 CSFs identified as a means of developing a CONQUAS management framework for optimum workmanship quality management. These literature findings are then tested through a survey questionnaire and supported by three interviews and one case study conducted. The results show that while the CSFs identified are known tenets of quality, they are still not being followed. Thus, it is important that the Building and Construction Authority (BCA) plays an important role to implement the proposed CONQUAS management framework on a national level to compel all contractors to adopt this model on a full scale and reap the benefits of being an enthusiast of adhering to the CONQUAS workflow.

Keywords Construction quality assessment system • Critical success factors • Project quality management • Singapore

Chapter 1
Introduction

1.1 Background

Quality has been a key issue in the construction industry since the late 1980s, with 37 % of all construction projects reporting major defects (Sullivan 2010). Murray (1993) commented that the low level of quality may be attributable to the fact that the industry was impervious to modern change and was structured as if nothing had changed in the last 50 years. In fact, the construction industry consists of numerous parties, each of which has a role to play in ensuring the quality of the product. The poor performance of one party will affect the performance of the next party. In addition, excessive changes to the details of the design of a project are typical throughout the construction process (Koehn and Regmi 1990). Quality performance is thus difficult to ensure. Poor performance will lead to disputes and adversarial relations between the parties which will again put future performance at risk, thus forming a vicious cycle of poor quality performance (Kanji and Wong 1998).

As a result, the construction industry has become inundated with serious problems in quality standards and requires the successful implementation of a quality management system (QMS) to deliver a consistent quality product and a platform for continual improvement. In Singapore, a firm-industry-national framework, namely the construction quality assessment system or CONQUAS and the International Organisation for Standardisation (ISO) 9000 were implemented by the Quality Development Unit of the Construction Industry Development Board, the predecessor of the building and construction authority (BCA), to improve quality performance.

1.1.1 Quality Management System

According to Crosby (1979), quality management is a systematic way of guaranteeing that organised activities happen the way they are planned. It is a management discipline concerned with preventing problems from occurring by creating the

S. P. Low and J. Ong, *Project Quality Management*, DOI: 10.1007/978-981-287-074-2_1,

attitudes and controls that make prevention possible (Saarinen and Hobel 1990). There are various QMSs which approach an organisation through different routes with the same goal in mind, that is, to achieve and sustain a high-quality output by conforming to requirements and meeting customer satisfaction requirements (Sullivan 2010). In fact, many construction-related firms have been implementing several QMSs to provide assurance that they can meet client's requirements, sustain their competitive advantage and, most importantly, manage quality problems.

The development and implementation of a sound QMS for use in the construction industry is indeed a necessity and not an option (Ramsey 1984). QMSs have to be directed and controlled with excellence so that high-quality buildings can be achieved. However, the quality of the implementation of these QMSs in individual construction companies is a matter of some debate (Conchúir 2011). Researchers claimed that the primary reason for certain companies to employ QMSs is simply to satisfy the mandatory requirement of the client rather than taking the full advantage of the QMSs to enhance their practices on a continuous basis (Ng 2005). This phenomenon is worrying as it is vital that all contractors are sincerely committed to apply practical strategies throughout planning, design and construction to achieve quality as outlined in the Project Management Body of Knowledge.

1.1.2 International Organisation for Standardisation 9000

Among various QMSs, ISO 9000 certification has been widely adopted by the construction industry of many countries (Ng 2005). The ISO 9000 series was launched to ensure quality standards are built into the operations to achieve consistency in the end product, and many contractors in Singapore have been accredited (Ofori et al. 2002). Moreover, the Singapore government has made it mandatory for larger construction and consultancy firms to achieve ISO 9000 certification as a pre-requisite for public sector projects. As of December 2012, 125 contractors, 8 public agencies and 47 consultants have been BCA ISO 9000 certified, while there are 12 property services and project management firms as well as 7 suppliers who are BCA ISO 9000 and Japan Quality Assurance Organisation certified in Singapore as compared to the extremely low figures when the scheme first started in 1991 (BCA 2012d). This shows that the ISO 9000 standards are beginning to play an essential role for contractors to have constant improvement in their processes, products and services, so much so that many contractors cannot choose to ignore the powerful influence of quality management.

Nevertheless, the ISO 9000 series only provides the foundation of a QMS. The only way to know whether a company is improving its overall quality is to measure periodically and compare the results after implementing any process or system with the historical results (Robert and Linda 2000). The way of measuring has to be as objective as possible to avoid subjective and confusing information that could lead to misinterpretations. A good measuring and fairly accurate system developed in Singapore by BCA is CONQUAS.

1.1.3 Construction Quality Assessment System

CONQUAS is used to measure the level of quality achieved in a completed building project using numerical scores. As part of the overall QMS, it provides a trusted and comprehensive assessment system to validate the contractor's workmanship excellence. It provides technical specifications for contractors to understand the required quality standards. Today, CONQUAS assessment is compulsory for all public sector projects with contract sum above S$5 million under the bonus scheme for construction quality (BSCQ) and private projects with CONQUAS requirement under individual contract agreement. It is voluntary for all other new building projects. BSCQ is set up to promote the upgrading of workmanship in the construction industry, in other words, to encourage contractors to achieve higher CONQUAS scores. Contractors will be paid a bonus for government projects or based on their individual agreements for private projects if their quality of workmanship exceeds a stipulated score. However, they will be penalised if their quality workmanship is poor. Expectations of quality standards by owners are ever-increasing, and hence, securing good CONQUAS scores should be a priority for all contractors to meet the changing demands of the construction industry.

1.2 Research Problem and Hypothesis

The demand for better workmanship of contractors has become more important in recent years (Griffith 2011). Firstly, owners and developers are better informed of good construction practices since the government stepped up stringent measures against building defects which put them in a better position to bargain and demand for quality than in the past. Secondly, the influx of reputable foreign contractors has meant a more competitive market environment and generated a need for the contractors to differentiate themselves by the workmanship quality which they can deliver (Oswald and Burati 1992). Thirdly, as local contractors are reaching out to foreign construction markets, the obligation to achieve good quality is becoming more compelling in order for them to improve their prospects abroad (Rommel 1996).

According to BCA (2012a, b), the average CONQUAS score for buildings in general has risen to 85.7 in 2011 compared to 76.5 in 2002 during the ten-year period. However, it is observed that the top performers are always from the same firms and there are still many other firms whose CONQUAS scores have not improved significantly and the expected continuous improvement in construction quality has not been realised. Furthermore, it is identified that there are merely 20 % of projects with scores above 90.1 and they only occurred in the recent five-year period from 2008 to 2012 (Fig. 1.1). This shows that much effort are still needed to raise the workmanship quality of all the contractors in Singapore, to attain a "high" CONQUAS score which according to the Cambridge dictionary,

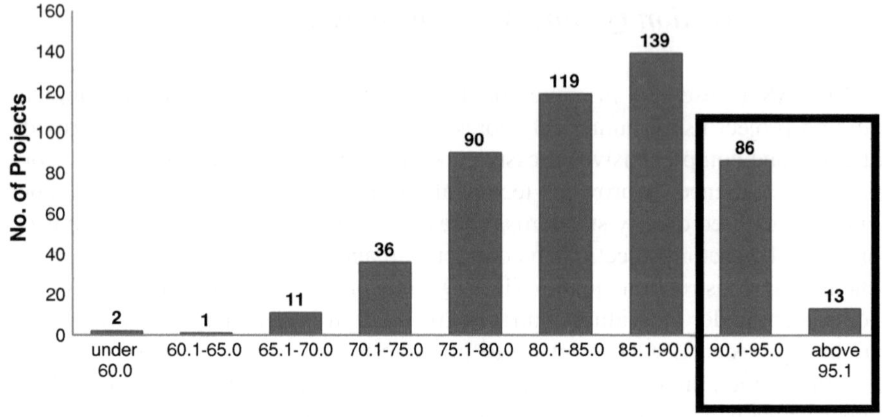

Fig. 1.1 Distribution chart of CONQUAS scores from 2008 to 2012. *Source* BCA (2013a, b)

means "greater than the usual level or amount". As Fig. 1.1 clearly presents that the construction industry has a low record in surpassing a score above 90, it will be taken to mean "high" CONQUAS scores in this research.

At the same time, these 99 projects with "high" CONQUAS scores (above 90.1) imply that there are certainly many ways in which a construction firm can achieve high CONQUAS scores. However, many projects do not seem to have formulated a comprehensive CONQUAS management program which is instrumental in achieving high CONQUAS score and ensuring that projects are run smoothly on site. On top of that, the variability of CONQUAS scores among the firms leads to the following seven sets of hypotheses about the critical success factors (CSFs) which will be subjected to testing and subsequent acceptance or rejection, following the analysis of the research results in Chap. 6.

Null Hypothesis, $H_{1.0}$: The more important CSFs do not show greater influence on the CONQUAS score.

Alternate Hypothesis, $H_{1.1}$: The more important the CSF is, the greater the influence it has on the CONQUAS score than CSFs which are not as important.

Null Hypothesis, $H_{2.0}$: The higher adoption rates of the CSFs do not show higher CONQUAS score.

Alternate Hypothesis, $H_{2.1}$: The higher the adoption of the CSFs, the greater the resultant CONQUAS score will be than when the CSFs are not adopted.

Null Hypothesis, $H_{3.0}$: The extent of variation of the usage level of the CSFs cannot be attributable to the variation in its importance rating.

Alternate Hypothesis, $H_{3.1}$: The extent of variation of the usage level of the CSFs can be attributable to the variation in its importance rating.

Null Hypothesis, $H_{4.0}$: The importance level of each of the CSFs does not correlate to the importance level of other CSFs.

Alternate Hypothesis, $H_{4.1}$: The importance level of each of the CSFs is correlated to the importance level of other CSFs.

Null Hypothesis, $H_{5.0}$: The extent of adoption of each of the CSFs does not correlate to the extent of adoption of other CSFs.
Alternate Hypothesis, $H_{5.1}$: The extent of adoption of each of the CSFs is correlated to the extent of adoption of other CSFs.

Null Hypothesis, $H_{6.0}$: The responses received from A1 contractors do not differ from A2 contractors with regard to the importance level of the CSFs.
Alternate Hypothesis, $H_{6.1}$: The responses received from A1 contractors differ from A2 contractors with regard to the importance of the CSFs.

Null Hypothesis, $H_{7.0}$: The responses received from A1 contractors do not differ from A2 contractors with regard to the extent of adoption of the CSFs.
Alternate Hypothesis, $H_{7.1}$: The responses received from A1 contractors differ from A2 contractors with regard to the extent of adoption of the CSFs.

1.3 Research Aim and Objectives

Although studies (Low et al. 1999; Ong 1997) have provided suggestions for construction firms to consider to attaining higher CONQUAS scores; at best, a prescribed list of items was proposed but they have not been evaluated and quantified in detail (Low 2001a, b). Furthermore, the study by Calingo et al. (1995) stated that little has been written on the nature and extent of journey of contractors towards achieving good workmanship quality in Singapore.

Hence, this research aims to bridge the knowledge gap by identifying the CSFs for achieving high CONQUAS scores by contractors from the time the project is awarded to the time the project is completed. CSFs are the limited number of areas in which results, if they are satisfactory, will ensure successful competitive performance for the organisation (Leidecker and Bruno 1984). It should be noted that CONQUAS does not measure the quality of building design, the materials specified nor the level of maintenance found in a building (Low 1993). It also does not cover latent defects that may appear after handing over as such defects cannot be foreseen or visible during CONQUAS assessment. Therefore, the CSFs that will be covered only involve the few key areas relating to construction workmanship quality which must be done right the first time according to the specifications and requirements of the consultants. The CSFs are those areas in which contractors must excel in order to be successful towards CONQUAS management.

Furthermore, it is important to not only evaluate the achievement of high CONQUAS scores by the CSFs during the project's construction phase, but perhaps to consider other influences from the various stakeholders of the project as well as corporate-wide performance and overall QMS of the company. Hence, to accomplish the aim of this research,

1. The whole CONQUAS process (Fig. 1.2) will be examined thoroughly to present the idea of CONQUAS and its assessment approach which have given rise to a certain quality trend.

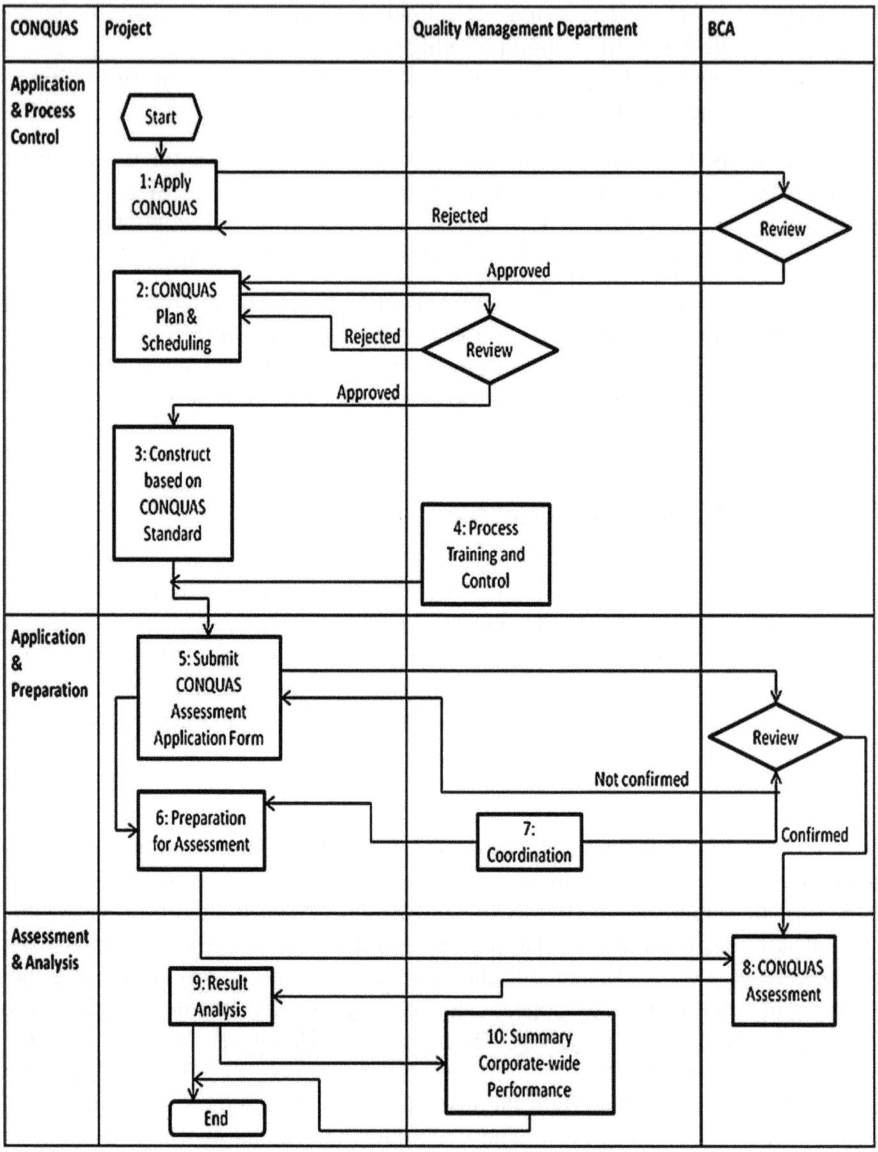

Fig. 1.2 CONQUAS flowchart. *Source* BCA Academy (2012c)

2. The QMS will then be investigated to explore its application to managing CONQUAS effectively.
3. Finally, CSFs for achieving high CONQUAS scores will be identified from secondary sources and they will then be verified from primary sources in order to develop a CONQUAS management framework for contractors to adopt in their future projects.

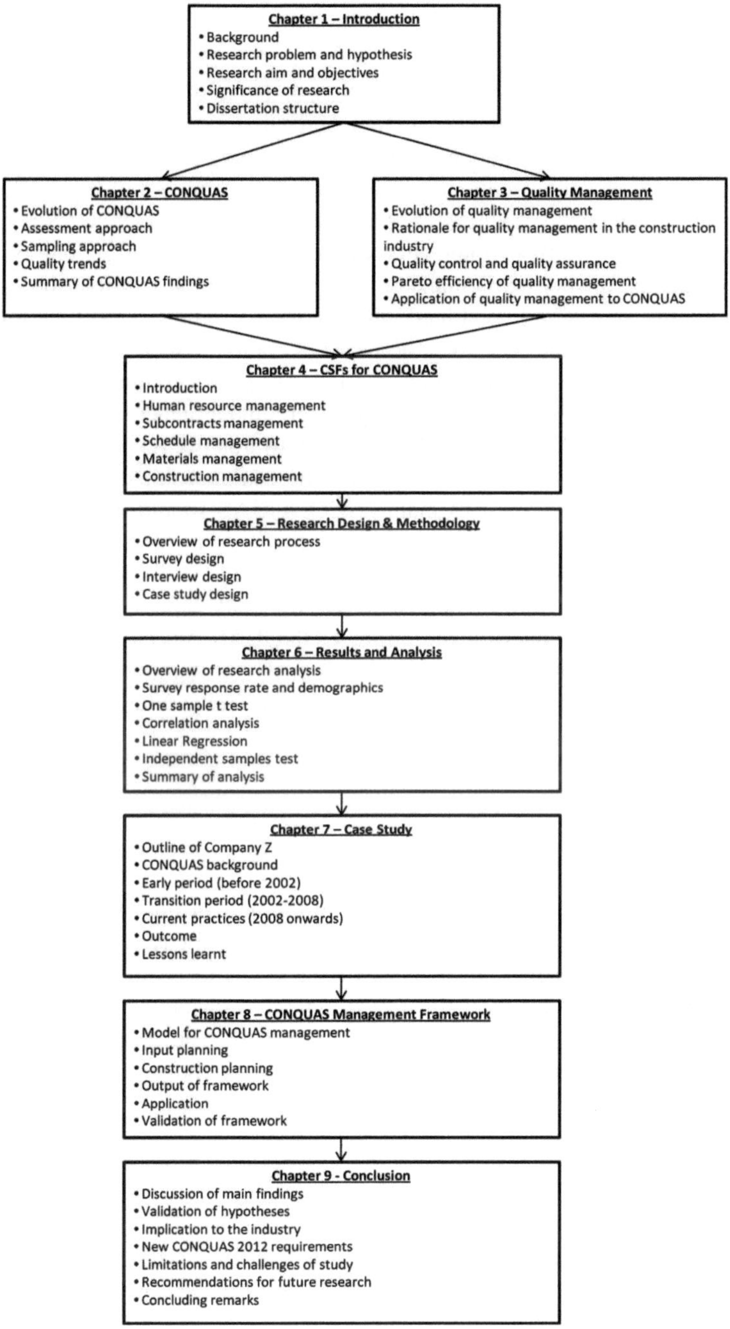

Fig. 1.3 Structure of book

1.4 Significance of Research

Having stressed the need to offer a clearer perspective to the successful achievement of high CONQUAS scores, this research endeavours to embark on the subject matter of overcoming existing challenges with the incorporation of CONQUAS in projects. This is significant as contractors with proven track record of high CONQUAS score projects have higher chances to secure new contracts. CONQUAS is also part of the Price Quality Method framework where CONQUAS score is one of the criteria to assess the quality performance of contractors which will then be translated into quantitative scores and combined with the price scores, to select the most suitable firm that provides the best offer for award of contract. Such greater emphasis on achieving high CONQUAS scores not only helps to gain a competitive advantage at the tendering stage, it also means monetary reward and a higher possibility of securing the prestigious BCA construction excellence awards (CEA) since CONQUAS is a major part of the evaluation criteria to assess the workmanship effort of the project.

Looking forward, the quest to manage CONQUAS should not be undermined to further improve overall quality standards and meet increasing developers and end-users expectation for better quality buildings. This will certainly bring about a progression towards partnership with clients and subcontractors for future projects. Moreover, conquering the entire CONQUAS process is a sign that the critical course of action was executed assiduously throughout the project which signifies the achievement of project targets as well as the attainment of organisational goals (CII and TQM Task Force 1994).

1.5 Book Structure

This book is divided into nine chapters as presented in Fig. 1.3. The first three chapters are a review of construction quality with focus on the CONQUAS scheme. Chapter 4 presents an analysis of the CSFs to be integrated into every construction project. With that, the following three chapters conduct a rigorous three-tier field investigation of the CSFs identified. The proposed CONQUAS management framework will then be discussed. Lastly, the main findings and validation of the seven sets of hypotheses will be summarised.

References

Building and Construction Authority. (2012a). Achieve better quality and higher productivity through conquas (8th ed.). Retrieved November 1, 2012 from http://www.bca.gov.sg/Professionals/IQUAS/conquas8.html

Building and Construction Authority. (2012b). Construction excellence award. Retrieved September 27, 2012 from http://www.bca.gov.sg/awards/constructionexcellence/construction_excellence_awards.html

Building and Construction Authority Academy. (2012c). Quality and productivity seminar 2012. Retrieved November 1, 2012 from http://www.bcaa.edu.sg/QPS2012.aspx

Building and Construction Authority. (2012d). BCA ISO 9000 Certification Scheme. Retrieved October 24, 2012 from http://www.bca.gov.sg/professionals/iquas/ISO.html

Building and Construction Authority. (2013a). BCA directory of registered contractors and licensed builders. Retrieved January 3, 2013 from http://www.bcadirectory.sg/search_result.php?workhead=MjA=&subworkhead=NA==

Building and Construction Authority. (2013b). Information on construction quality. Retrieved January 5, 2013 from http://www.bca.gov.sg/professionals/iquas/bqphomepage.asp

Calingo, L. M. R., Leong, Y. M., Chia, M. P., & Mohamed, H. (1995). Achieving total quality management through ISO 9000: a research note. *Accounting and Business review*, *2*(1), 173–186.

Conchúir, D. Ó. (2011). *Overview of the PMBOK® guide: Short cuts for PMP® certification*. Berlin: Springer.

Construction Industry Institute and TQM Task Force. (1994). Implementing TQM in engineering and construction. US: University of Texas at Austin.

Crosby, P. B. (1979). *Quality is free: The art of making quality certain* (Vol. 94). New York: McGraw-Hill.

Griffith, A. (2011). Integrated management systems for construction: Quality. Environment and Safety. Sheffield Hallam university, UK: Trans-Atlantic Publications.

Kanji, G. K., & Wong, A. (1998). Quality culture in the construction industry. *Total Quality Management, 9*(4–5), 133–140.

Koehn, E., & Regmi, D. (1990). Quality in constructed projects: International firms and developing countries. *Journal of Professional Issues in Engineering, 116*(4), 388–396. doi:10.1061/(ASCE)1052-3928(1990)116:4(388).

Leidecker, J. K., & Bruno, A. V. (1984). Identifying and using critical success factors. *Long Range Planning, 17*(1), 23–32. doi:10.1016/0024-6301(84)90163-8.

Low, S. P. (1993). The conceptual relationship between construction quality and economic development. *International Journal of Quality & Reliability Management, 10*(2), 18–30.

Low, S. P. (2001a). Quantifying the relationships between buildability, structural quality and productivity in construction. *Structural Survey, 19*(2), 106–112.

Low, S. P. (2001b). Singapore's construction industry: Achieving higher quality standards. *The Singapore Exchange Journal*, Pulses(June), pp. 40–43.

Low, S. P., Tan, B. K., & Ang, A. A. L. (1999). Effectiveness of ISO 9000 in raising construction quality standards: Some empirical evidence using CONQUAS scores. *Structural Survey, 17*(2), 89–108.

Murray, P. (1993). *A fantastic journey: The life and literature of Lafcadio Hearn*. UK: Psychology Press.

Ng, S. T. (2005). Performance of engineering consultants in ISO 9000-based quality management systems implementation. *Engineering, Construction and Architectural Management, 12*(6), 519–532.

Ofori, G., Gang, G., & Briffett, C. (2002). Implementing environmental management systems in construction: Lessons from quality systems. *Building and Environment, 37*(12), 1397–1407. doi:10.1016/S0360-1323(01)00115-9.

Ong, B. H. (1997). Improving quality performance in construction using a total quality management model. (M. Eng.), University of Malaya, Kuala Lumpur.

Oswald, T. H., & Burati, J. L. (1992). Guidelines for implementing total quality management in the engineering and construction industry. Bureau of Engineering Research, University of Texas at Austin, Austin.

Ramsey, T. (1984). Quality control "a necessity not an option". *Journal of Construction Engineering and Management, 110*(4), 513–517. doi:10.1061/(ASCE)0733-9364(1984)110:4(513).

Robert, A. O., & Linda, L. B. (2000). An integrated view of project and quality management for project-based organizations. *International Journal of Quality and Reliability Management, 17*(4), 351–363.

Rommel, G. (1996). *Quality pays*. New York: Macmillan Press.

Saarinen, A., & Hobel, M. (1990). Setting and meeting requirements for quality. *Journal of Management in Engineering, 6*(2), 177–185. doi:10.1061/(ASCE)9742-597X(1990)6:2(177).

Sullivan, K. (2010). Quality management programs in the construction industry: Best value compared with other methodologies. *Journal of Management in Engineering, 27*(4), 210–219.

Chapter 2
Construction Quality Assessment System

2.1 Evolution of CONQUAS

2.1.1 History and Development

In 1989, the first edition of CONQUAS was introduced to evaluate the quality performance of building contractors in the public sector (Tang et al. 2005). Subsequently, CONQUAS was applied to the superstructure works of private building projects in 1991 as well as development on sites sold by the Housing and Development Board and the Urban Redevelopment Authority and civil engineering works construction in 1993 as a way to assure quality even in other sectors. In the fifth edition launched in 1998, known as CONQUAS 21, the assessment of Mechanical and Electrical (M&E) Works was included to replace the External Works component to make CONQUAS scoring more accurate and customer-oriented (Chiang et al. 2005). Industry concerns and end-user feedback continued to shape CONQUAS 21 BCA (2005). After a review focusing on latent defects, the sixth edition launched in 2005 introduced the wet-area water-tightness testing and in-process inspection for internal wet-area waterproofing works to ensure better quality assurance and higher CONQUAS scores. Following that, the seventh edition became applicable in 2008 where the defect level weightages for internal finishes assessment are raised for a more accurate reflection of homeowner priorities on defects. Hence, this study will be based on the seventh edition, which can be found from the following web link:

http://www.bca.gov.sg/professionals/iquas/others/CONQUAS_7edit.pdf

However, it should be noted that the latest eighth edition, which was launched on 31st October 2012 (mid-way during this study), can be found in:

http://www.bca.gov.sg/professionals/iquas/others/CONQUAS8.pdf

In a nutshell, CONQUAS is reviewed periodically due to changes and improvements in processes, technology, strategies and methods of construction, which are continuously evolving in the industry. Hence, it is necessary to constantly align the CONQUAS standard with industry trends to keep it current and relevant.

S. P. Low and J. Ong, *Project Quality Management*, DOI: 10.1007/978-981-287-074-2_2,

Table 2.1 Bonus and discount threshold score from 1/4/2012 to 31/3/2013

Building category	Discount threshold score	Bonus threshold score
Residential	82.6	88.6
Commercial	82.0	88.0
Institution	80.0	86.0
Industrial/others	79.1	85.1

Source BCA (2012)

Table 2.2 Merit and default points

CONQUAS score (%)	Merit/default points
>(A + 15)	5 merit points
(A + 12.1)–(A + 15)	4 merit points
(A + 9.1)–(A + 12)	3 merit points
(A + 6.1)–(A + 9)	2 merit points
(A + 3.1)–(A + 6)	1 merit point
(A − 3)–(A + 3)	Nil
(A − 6)–(A − 3.1)	1 default point
(A − 9)–(A − 6.1)	2 default points
(A − 12)–(A − 9.1)	3 default points
(A − 15)–(A − 12.1)	4 default points
(A − 20)–(A − 15.1)	5 default points
< (A − 20)	10 default points (debarment to be considered)

Note A is the average CONQUAS score for the particular building category

Source BCA (2012)

2.1.2 Bonus Scheme for Construction Quality

CONQUAS 21 was launched together with the BSCQ whereby a contractor of a public project would have a 0.2 % bonus or discount of the effective contract sum for every point scored above or below the bonus or discount threshold score. The bonus or discount threshold scores are set at three points above and three points below the previous 24-month average CONQUAS score for the relevant building category as shown in Table 2.1. This would give them a preferential advantage of up to 3 % of the effective contract sum or S$2 million, whichever is lower, over their competitors when tendering for government projects. As a result, contractors have become more conscious of quality as those who performed poorly would be penalised with disincentives (BCA 2009).

Furthermore, based on the latest five contracts, when a contractor accumulated CONQUAS default points as explained in Table 2.2, a price-loading of 0.2 % for each CONQUAS default point, subject to a maximum of S$2 million, would be applied against any tender proposal by the contractor in the evaluation of tender. An example of how this price-loading affects the tender evaluation process is shown in Table 2.3. Apart from that, once the contractor accumulates more than

Table 2.3 Example of application of price-loading

Firm	Total tender sum (S$m) and accumulated default points	Effective tender sum (%)	Price-loading factor (%) (0.2 % for each default point)	Computation of price-loading (S$m)	Applicable price-loading (S$m)	Effective tender value (S$m)
	(a)	(b)	(c)	(d) = (a) × (b) × (c)	(e) = (d) subject to S$2 m maximum	(f) = (a) + (e)
A	301.0 (2 default points)	90	0.4	1.0836	1.0836	302.0836
B	300.0 (4 default points)	90	0.8	2.16	2.0	302.0
C	301.3 (1 default point)	90	0.2	0.54234	0.54234	301.84234

Note In the above example, based on the effective tender value, Firm C would be awarded the tender at his tendered sum of S$301.3 million, even though Firm A and B have tendered lower actual prices of S$301 million and S$300 million, respectively

Source BCA (2012)

five CONQUAS default points, it will be downgraded by one financial grade for up to a period of twelve months. Alternatively, debarment will be recommended if the contractor accumulates ten or more default points.

Therefore, contractors who want to tender and win in a public project have to have a good track record of high CONQUAS scores. With that, chances of any penalties and disincentives will be minimised as well. To a large extent, relying on the BSCQ to enforce workmanship in contractors does assure that the buildings delivered will be of a certain quality (Mohammed and Tan 2001). Moreover, this policy has been welcomed by contractors as it is deemed to be an effective policy introduced by the government to drive the quality standards in the industry (Mohammed and Tan 2001). Coupled with the Construction 21—Reinventing Construction's vision of transforming the construction industry to be a world-class builder in the knowledge age (Construction 21 Steering Committee 1999), it is certainly important for contractors to devise a list of CSFs to the achievement of high CONQUAS score so that the level of built quality can be delivered with greater assurance.

2.1.3 Introduction of Quality Mark Scheme

The evolvement of CONQUAS is supported with the launch of the quality mark (QM) for Good Workmanship Scheme, which is issued to individual apartment unit for new residential projects. QM is mainly evolved from the CONQUAS Internal Finishes Quality standards. Each unit has to achieve a minimum of 80 points, and there should not be any leakages detected during the water ponding test and window water-tightness test (optional). This is to help and encourage developers meet the rising expectation of Singaporeans for better consistent quality homes. Although QM is completely voluntary, its take-up rate has increased steadily from 28 % in 2006 to 56 % in 2009 for the purpose of enhancing both developer's and contractor's branding (BCA Academy 2012c). The QM average unit score performance has also improved from 80.8 in 2006 to 83.9 in 2009, and this has been found to be correlated to its attainment of higher CONQUAS score as compared with non-QM private residential projects as shown in Fig. 2.1.

Nonetheless, there is a slight difference between CONQUAS and QM as depicted in Table 2.4. QM certification is for individual unit that provides a better indication of internal quality level only, while CONQUAS is a certification of the overall project quality which may actually varies from unit to unit. In fact, participation in QM has helped to propel Singapore's construction industry workmanship standards to a greater height, raising the percentage of building projects achieving a CONQUAS score above the benchmark. However, this has only been successful to private residential project (QM certified), while other types of projects (non-QM) are still far from achieving a CONQUAS score above the norm (BCA Academy 2012c). Even so, no matter whether projects are QM certified or not, it is vital that contractors have the skills and capability to manage CONQUAS by way of formulating a set of CSFs for achieving high CONQUAS scores.

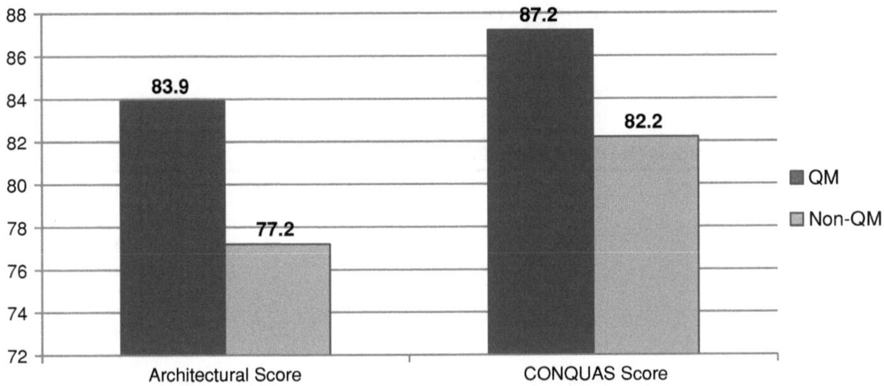

Fig. 2.1 Average architectural and CONQUAS score for QM and non-QM private residential projects from year 2006 to 2009. *Source* BCA Academy (2012c)

Table 2.4 Difference between CONQUAS and QM scheme

Criteria	CONQUAS	Quality mark
Applicant	Main contractor	Developer
Assessment fee	Based on gross floor area (GFA) of project	Based on unit rate for: Condominium development Landed housing
Scope of assessment	Structural works	Internal finish works
	Architectural works	Waterproofing test to bathrooms
	Internal finishes	Random in-process inspection on key trades:
	External wall/works	Waterproofing works
	Window water-tightness test	Marble/tiling works
	Pull-off test for wall tiles	Timber flooring works
	Material and functional tests	Window installation
	M&E works	
Assessment sampling/ approach	Sampling approach	Internal finish of all units
	Samples worked out based on GFA of project	Water ponding test of all bathrooms found in units
	1st time right approach	Allow re-score
Assessment outcome	CONQUAS certificate for project with score reflected	Quality mark certificate for every individual unit that meets the stipulated standard
	Two certificates issued: developer	Individual unit score not reflected on certificate
	Main contractor	Certificates are issued to developer only

Source BCA Academy (2012)

Table 2.5 CONQUAS score weightage system

Components	CAT A: commercial, industrial, institution and others (%)	CAT B1: commercial, industrial institution and others (%)	CAT B2: private housing (%)	CAT C: public housing (%)	CAT D: landed housing (%)
Structural works	25	30	25	35	30
Architectural works	55	60	65	60	65
M&E works	20	10	10	5	5
CONQUAS score	100	100	100	100	100

Note In general, projects with central cooling system having cooling tower, chiller system, etc. are classified under CAT A. Otherwise, it will be classified under CAT B1

Source BCA (2008)

2.2 Assessment Approach

The assessment is divided into three parts–structural works, architectural works and M&E works–with different weightages for each building category. This is aimed at making the CONQUAS score objective in representing the quality of a building to reflect the approximate cost ratio of each component in the various building types and their aesthetic consideration as shown in Table 2.5. The quality of workmanship will be assessed throughout the construction process for structural and M&E works. On the other hand, the workmanship quality for architectural works is assessed on buildings completed between one to three years which give rise to a higher chance of subjecting to lower CONQUAS scores as the workmanship quality may worsen upon being occupied by users. While the lower limit of one year helps to ensure that faults, if any, can be detected, the upper limit of three years will ensure that the building concerned can still be regarded as a relatively new development. Overall, CONQUAS provides a common objective and measureable platform for quantifying the quality standards within specific time and cost limits and in the process, raises the level of quality in construction BCA (2005).

Generally, the projects are assessed through site inspections, tests on materials and functional performance of selected services and installations, where the workmanship is evaluated and scored objectively by trained BCA assessors using standard score sheets BCA (2008). Figure 2.2 shows an example of an internal finishes assessment in the principal location. The required samples are chosen according to the ratio set out in the four categories of buildings for the principal (e.g. halls and rooms), circulation (e.g. stairs, corridors and lift lobbies) and service (e.g. kitchen, toilets and plant room) areas, respectively. This is vital as all three locations will ultimately affect the long-term performance of the building. Most importantly, scoring will only be done once and rectification or any correction carried out thereafter will not be re-assessed to encourage the approach of "doing things right the first time".

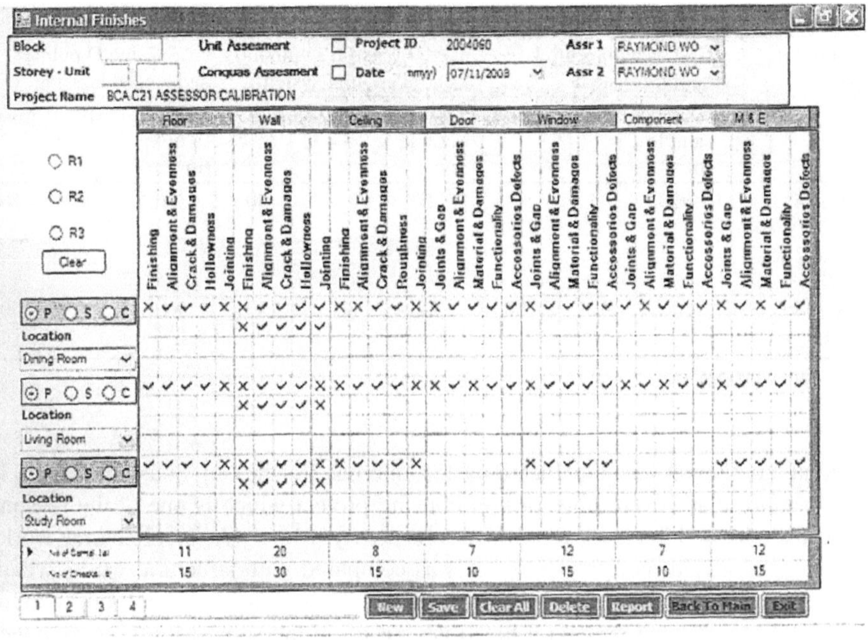

Fig. 2.2 Example of CONQUAS assessment internal finishes score sheet. *Source* BCA Academy (2012)

Table 2.6 Number of samples required for structural and architectural works

Components	CAT A and B1 commercial, industrial, institution and others			CAT B2 and D private and landed housing			CAT C public housing		
	GFA per sample (m^2)	Min.	Max.	GFA per sample (m^2)	Min.	Max.	GFA per sample (m^2)	Min.	Max.
Structural elements	500	30	150	1,500	30	50	1,500	30	50
Architectural internal finishes				70	30	800	70	30	600

Source BCA (2008)

2.3 Sampling Approach

The number of samples that is required to be prepared is believed to have affected the CONQUAS score. From Table 2.6, it is shown that the number of structural sample required for housing projects is thrice lower than non-housing projects. This is the main reason which accounts for the fact that structural works of housing projects tend to perform better than non-housing projects (Corenet 2012). As a result,

Table 2.7 Number of samples required for M&E works

Components	CAT A with central cooling system			CAT B1 and B2 without central cooling system			CAT C and D public and landed housing		
	GFA per sample (m^2)	Min.	Max.	GFA per sample (m^2)	Min.	Max.	GFA per sample (m^2)	Min.	Max.
Electrical, ACMV, fire protection, S&P	1,000	35	70	1,500	25	50	3,500	10	20
Basic M&E fittings	500	30	150	500	30	150	500	30	150

Source BCA (2008)

the total number of sample locations for internal finishes of housing project is notably more than seven times of non-housing project which is one of the reasons that causes its architectural works to score poorly. Moreover, architectural works have to be assessed on a free-look basis as compared with structural works where sample locations have to be planned beforehand. Thus, contractors will put in more effort in these planned structural locations but to do the same for all the architectural works would be tough.

Next, although M&E works are also randomly assessed, they are still able to score relatively well (Corenet 2012) due to the fact that M&E works have a relatively much lower number of sample locations required as compared with structural or architectural elements as shown in Tables 2.6 and 2.7. With that, it is suggested that to a certain extent, the CONQUAS sampling approach is deemed to have unwittingly manipulated the quality trend results.

Overall, the number of sample locations required for architectural works is the most out of the three components which signifies that more attention has to be paid to ensure that all architectural sample locations are thoroughly ready for CONQUAS assessment. Moreover, architectural works accounts for at least 55–65 % of the CONQUAS score depending on the category of building it belongs to, acting as a barrier to achieving a high CONQUAS score. Hence, it is important that contractors take note of the quality trends in each of the three components so as to allocate resources more appropriately and tailor a set of CSFs to overcome this shortcoming of CONQUAS assessment and cultivate a "first time right" mindset as well as to achieve high CONQUAS scores.

2.4 Quality Trends

In order to do so, industry players can look at the information on construction quality (IQUAS) website which provides a vast repository of CONQUAS assessment data to benchmark their performance on workmanship quality against the industry standards. A noticeable defect trend data on (Corenet 2012) found that the architecture works component has the highest percentage of non-compliance ever since CONQUAS was launched. This is further supported as the average score of architectural works is only 73.9 as compared with 86.6 and 86.5 achieved in the structural and M&E components, respectively (Corenet 2012).

In view of this, BCA developed a CONQUAS 21 Enhancement Series called the good industry practices (GIP) guides which aims to share with contractors good industry work practices adopted by contractors and practitioners who have been able to consistently deliver high-quality work, in other words, high CONQUAS scores. These guides, also available in the CONQUAS application, serve to help improve the contractor's quality standards with focus on the architectural aspect. The following twelve titles have been released since 2003:

- Ceramic Tiling (Second Edition)
- Marble and Granite Finishes (Second Edition)
- Waterproofing for Internal Wet Areas (Second Edition)
- Painting (Second Edition)
- Waterproofing for External Wall
- Timber Flooring
- Aluminium Window
- Timber Doors
- Wardrobes and Kitchen Cabinets
- Precast Concrete Elements
- Design and Material Selection for Quality—Volume 1
- Design and Material Selection for Quality—Volume 2

In particular, the use of dry walls on average has lead to a higher score of 87.2 % compared with 76.7 % achieved using conventional brick wall or reinforced concrete wall (BCA 2012b). In spite of this, architectural works has only achieved an overall improvement rate of 52 % from 2001 to 2011 as compared with the improvement rate of 84 and 68 % achieved in structural works and M&E works, respectively (BCA 2012b).

As a whole, the introduction of GIP is targeted at the upstream construction activities whereby selection of materials and design plans drawn up by the consultants is deemed to be of great consequence to the downstream activities of the contractors, playing a major role in influencing the workmanship quality on site. With that in mind, the following sections will highlight on the downstream workmanship quality trends for each assessment component.

2.4.1 Mechanical and Electrical Works

In this segment, 50 % of the score is for the M&E works on-site inspection, and the other 50 % is for the performance test assessment declared by the qualified personnel. It is observed that M&E works has the lowest number of defects non-compliance and a proportionately higher average score achieved than architectural works though slightly lower than structural works (Corenet 2012). Apart from the low number of sample locations required as mentioned, Griffith (2011) has recognised that there are much more national, regional and international standards including the Singapore Standards for M&E works and even structural components as compared with architectural components. This accounts for the relatively decent performance by M&E works as they have a reference point to understand and apply the correct system so as to attain the stipulated workmanship quality required.

2.4.2 Architectural Works

Besides the sampling and assessment approach that makes it more challenging to achieve high architectural scores, another reason for the poor performance of architectural works is largely due to the fact that they are the last trade to begin work. As the initial phase of the project is often delayed, architectural works have lesser time to complete subsequently (Fewings 2005). Consequently, the abundance of demanding interfacing works, trades and details in architectural works cannot be attended to attentively which caused workmanship standards to suffer, leading to an increase in percentage of non-compliance. In addition, during construction, it is evitable that labourers need to frequently access various areas to complete their work. Due to the lack of protection of the materials at the factory as well as during delivery at site and after installation, architectural trades, being the outer layer, are exposed to higher risk of deteriorating faster than how it should be subjected to wear and tear under normal operational conditions (Meier and Wyatt 2008). However, even though contractors play a major part in affecting the architectural scores, architects and owners cannot deny their responsibility as the method, system and technology chosen in consideration of their budget and needs also play a major role in affecting the quality of workmanship that contractors cannot control.

2.4.2.1 Internal Finishes

There are six elements assessed in internal finishes, and it is observed that the percentage of non-compliance achieved in each element have been rather stagnant since 2007 to 2011 as shown in Fig. 2.3. On top of that, the percentage of non-compliance of each and every element in 2011 is actually higher than the value in year 2000 which suggests that there has been no or little improvement and more has to be done to reduce the percentage of non-compliance for all elements. In particular, the jointing and gaps defect is present across all six elements with an exceptionally high percentage of non-compliance as compared with the other type

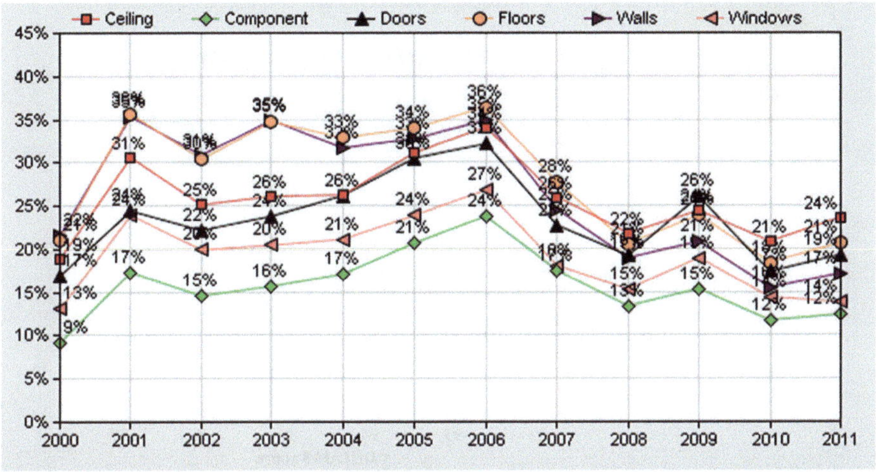

Fig. 2.3 Percentage of non-compliance for internal finishes from 2000 to 2011. *Source* (Corenet 2012)

of defects (Corenet 2012). Apparently, this seems to suggest that current measures have not been effective or rather, not utilised at all to reduce the number of non-compliance.

The GIP measures recommend that segregation of tiles based on production batch as well as the use of rectified (first-grade) tiles will assure a better workmanship quality. However, contractors may be concerned that more time and cost will be involved to achieve such workmanship accuracy (Chiang et al. 2005). Another GIP is the choice of using rebated door system with lift-off hinges whereby the doors are kept and will only be installed at a much later stage during the construction period to minimise damages as construction of other trades are still in progress. Alternatively, sub-frame door system is also encouraged compared with the traditional system. Similarly, although the costs of these unconventional door systems are much higher, they are easier and faster to install and more convenient to handle, which will lead to an easier means to achieve the required workmanship quality. Hence, contractors rather forgo such GIP and sacrifice the CONQUAS score in order to spend within budget and earn more profits, which is the priority of most contractors.

On the whole, the findings suggest that the use of such GIP is acceptable as long as it does not lead to great diminishing returns. This means that the achievement of high CONQUAS scores should not be at a significant expense of increased manpower and cost which is also the rationale behind the capping of CONQUAS score at 95 in the latest CONQUAS eighth edition, published in October 2012. This is supported by studies conducted by BCA (2012) which found that in order to increase the CONQUAS score from 95 to 97, just two points, requires a significant increase of 44 man-days as shown in Fig. 2.4.

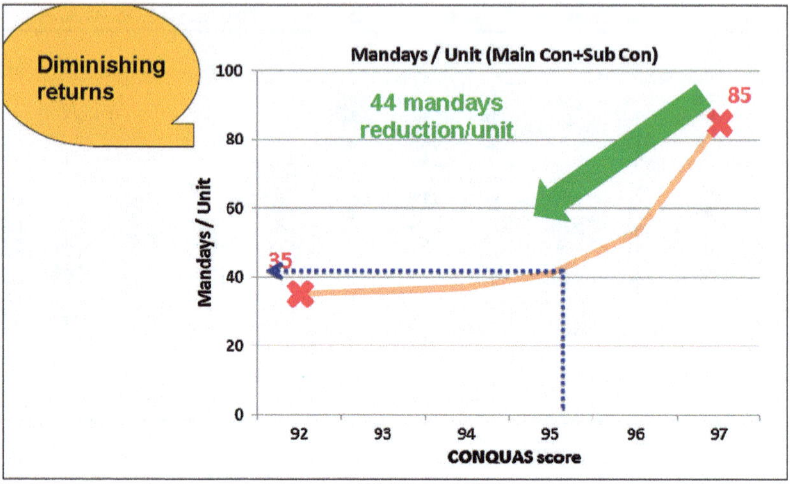

Fig. 2.4 Number of man-days versus CONQUAS score. *Source* BCA (2012)

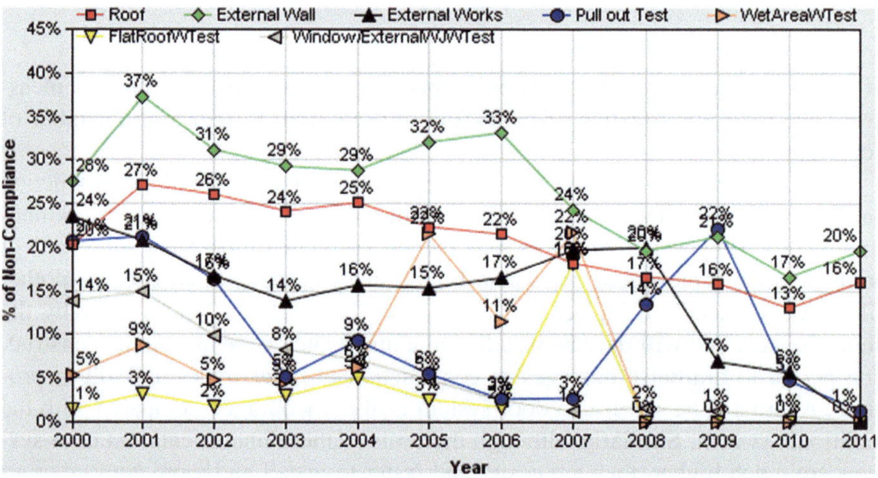

Fig. 2.5 Percentage of non-compliance for other architectural works from 2000 to 2011. *Source* (Corenet 2012)

2.4.2.2 Other Architectural Works

Although this segment has a much lower percentage of non-compliance as compared with the internal finishes segment and improvement has been passable (BCA 2012c), more can be done to further perfect the score of this segment so as to be on par with the almost perfect value achieved in the water-tightness test (WTT) elements from 2008 to 2011 as shown in Fig. 2.5. This may be due to the fact that self-testing is required, and hence, corrective actions can be done before the official test by BCA. In addition, BCA have to conduct an in-process

Table 2.8 Industry non-compliance average by assessment type

Type of defects	Industry non-compliance average (%)
Roof	22.66*
External wall	28.77*
External works	22.56*
Pull-Off test	11.29
Wet-area water-tightness test	8.73
Flat roof water-tightness test	3.25
Field window water-tightness test	5.93
Total	21.25

Source (Corenet 2012)

* The top three areas with a high percentage of non-compliance are indicated with an asterisk

inspection of the internal wet-area waterproofing process based on the approved method statement and shop drawings before actual works begins which contributes to the low level of non-compliance in the WTT. With that, BCA have decided to reduce the WTT weightage from ten points to nine points in the latest CONQUAS eighth edition so as to place more emphasis on better quality design and material choices.

Unlike internal finishes, the roof, external wall and external works are assessed after the temporary occupation permit is issued. Hence, contractors are advised to make an appointment for assessment as soon as possible due to the fact that these three elements are open areas and will be subjected to varying weather conditions. Such exposure will certainly affect the workmanship quality that was constructed in the first place which explains the industry trend of having a relatively high percentage of non-compliance in these three areas (indicated with an asterisk) as depicted in Table 2.8.

2.4.3 Structural Works

Structural works is deemed as the root cause of problem and any poor workmanship quality detected will affect subsequent trades, resulting in poorer workmanship quality of the end product (Hoonakker et al. 2010). The determination of nonconforming structural work is more difficult when the work has been covered by finish work or subsequent installations (Demkin 2008). Moreover, to fast-track construction, structural components are often poured at one go first without making openings for electrical services as the exact locations have not been confirmed yet during the early stage of construction (Fanella 2010). Demolishing parts of the structural component after that, will affect the workmanship quality of the final product which is also one of the reasons why these areas, are often not chosen to be part of the sample location. As mentioned earlier, samples for structural works are pre-planned by the main contractor, and this is an opportunity for them to set up a high level of workmanship standard for the assessment, which may be the reason why structural works performed the best out of the three components.

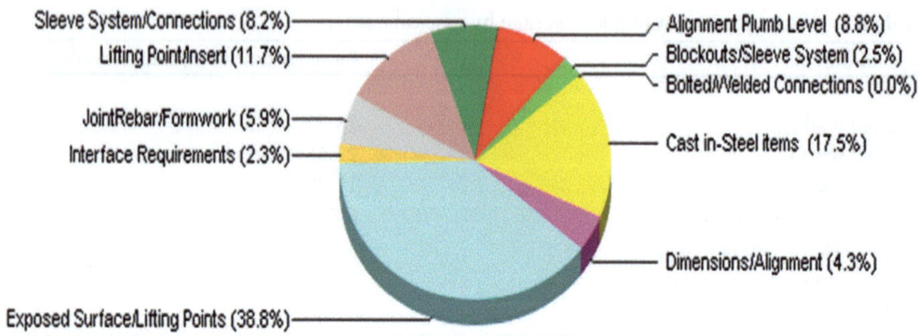

Fig. 2.6 Precast concrete defect distribution from 01/01/2000 to 30/04/2011. *Source* (Corenet 2012)

It is noted that precast concrete, prestressed concrete and structural steel work have relatively lower percentage of non-compliance (Corenet 2012) largely because these systems are produced under factory-controlled conditions (ACI 2008). In particular, the exposed surface criterion is a major contributor to the non-compliance for precast concrete as shown in Fig. 2.6. This is because of damages suffered due to lifting operations which is harder to accomplish especially with extreme site constraints and load limitations of the tower crane (Peurifoy et al. 2010). Nevertheless, it is still easier to control the workmanship quality of precast elements as compared with other types of structural components. Therefore, the use of precast elements is widely promoted as seen in the latest CONQUAS eighth edition where bonus points will be awarded when at least 65 % of toilets are prefabricated and if accreditated precasters are employed too. This is to reduce the need to deploy skilled manpower to carry out the finishing works, which are highly intensive on these components.

On the contrary, reinforced concrete structural system has the highest percentage of non-compliance as its workmanship quality demands more on the site conditions and skill level of the labourers which is harder to control (Meier and Wyatt 2008). Studies have found that the type of formwork and rebar chosen will affect the workmanship quality of the structural works greatly (Tattersall 1990). The two main types of rebar are mesh kind which is fixed in the factory, and the traditional method where rebars are tied manually on site. Similarly, factory-controlled mesh will be of better quality but cost is an issue (American Concrete Institute 2008). Hence, there is still a need for manual bar bending to reduce cost but its quality may not be as good as mesh rebars. Next, system formwork will ensure that a better quality finishing be achieved as compared with timber or metal formwork which tends to deteriorate faster when they are re-used to construct the next level (Oberlender and Peurifoy 2010). This is one of the reasons for a high percentage of non-compliance of the finished concrete, and Fig. 2.7 shows that the biggest source is from the exposed surface criteria where coarse aggregates and bulging are often detected due to the gaps in the formwork.

Additionally, the ultrasonic pulse velocity test for concrete uniformity has a very high percentage of non-compliance in the structural works component.

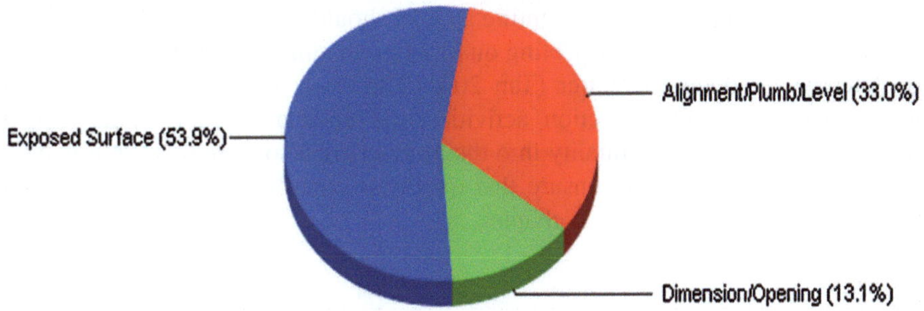

Fig. 2.7 Finished concrete—percentage defect distribution from 01/01/2000 to 30/04/2011. *Source* (Corenet 2012)

This is common in reinforced concrete structure due to the fact that continuous pour to each structural element from the same truck of the same concrete batch was not ensured due to poor estimation planning of the concrete required (ACI 2005). This is critical because if concrete of different batches are used; there will be higher chances of aggregation which compromises on the workmanship quality of the final building product.

2.5 Summary of CONQUAS Findings

To transform Singapore's building quality excellence, these common areas of non-compliance have to be eliminated. Among all three components, it is noticed that the products, systems and methods of construction chosen is critical in affecting the CONQUAS score. Moreover, the source of the major contributing factor to the high number of non-compliance actually relates to the design and materials selected. This implies that the specifications in the contract documents are influential to the contractor's effort to achieve high CONQUAS scores. Hence, in order to meet the minimum CONQUAS standards, these specifications could be drafted out based on the National Productivity and Quality Specifications which can be found electronically (eNPQS).

eNPQS aims to harmonise the industry building specifications and provide a standard platform for achieving greater efficiency and quality in the design and construction process. It is written with reference to recognised standards as well as the CONQUAS standards. This implies that having specifications drafted no less than the minimum criteria of eNPQS is an important role that the architect has to play to ensure that contractors are able to meet the CONQUAS requirements by complying with the contractual specifications (Lee et al. 2011). This also means that the contractual specifications should have incorporated the workmanship quality requirements to assure that the CONQUAS management workflow designed will at least be able to pull-off and achieve the minimum CONQUAS score. Moreover, submitting a quality control plan to the client is a criterion in the eNPQS.

Essentially, companies implementing a QMS should emphasise building quality into the product rather than inspecting quality into the finished product and removing defective products thereafter (Tan 2001).Therefore, besides the downstream quality inspection and correction activities, the upstream quality management planning activities to build quality into the product are also significant in influencing the CONQUAS score to ensure that quality is controlled as it should be and this will be presented in the next chapter.

References

Tang, S.-l., Ahmed, S. M., Aoieong, R. T., Poon, S. W. (2005). Construction quality management: Hong Kong University Press, Hongkong.

American Concrete Institute. (2005). The contractor's guide to quality concrete construction: American Society of Concrete Contractors.

American Concrete Institute. (2008). Building Code Requirements for Structural Concrete (ACI 318-08) and Commentary.

Building and Construction Authority. (2005). CONQUAS 21: The BCA Construction Quality Assessment System. Singapore.

Building and Construction Authority. (2008). CONQUAS: The BCA Construction Quality Assessment System. Singapore.

Building and Construction Authority. (2009). Quality and certification, CONQUAS: Frequently Asked Questions. Retrieved November 18, 2013, from http://www.bca.gov.sg/professionals/iquas/bqphomepage.asp

Building and Construction Authority. (2012a). Achieve better quality and higher productivity through CONQUAS 8th edition. Retrieved November 1, 2012, from http://www.bca.gov.sg/Professionals/IQUAS/conquas8.html.

Building and Construction Authority. (2012b). Construction excellence award. Retrieved September 27, 2012, from http://www.bca.gov.sg/awards/constructionexcellence/construction_excellence_awards.html.

Building and Construction Authority Academy. (2012c). Quality and productivity seminar 2012. Retrieved November 1, 2012, from http://www.bcaa.edu.sg/QPS2012.aspx.

Construction 21 Steering Committee (1999). *Re-Inventing Construction*. Ministry of Manpower and Ministry of National Development, Singapore: McGraw Hill Financial.

Corenet. (2012). *Information on construction quality*. Retrieved December 17, 2012, from http://www.corenet.gov.sg/iquas/

Chiang, Y. H., Raftery, J., Anson, M. J. (2005). The construction sector in the asian economies: Spon Press, New York.

Demkin, J. A. (2008). The architect's handbook of professional practice: Wiley.

Lee, D. E., Lim, T. K., & Arditi, D. (2011). An expert system for auditing quality management systems in construction. *Computer-Aided Civil and Infrastructure Engineering, 26*(8), 612–631.

Fanella, P. D. (2010). Reinforced concrete structures: analysis and design: McGraw-Hill.

Fewings, P. (2005). Construction project management: An Integrated Approach: Taylor & Francis, New York.

Griffith, A. (2011). Integrated management systems for construction: Quality, environment and safety: Trans-Atlantic Publications.

Hoonakker, P., Carayon, P., & Loushine, T. (2010). Barriers and benefits of quality management in the construction industry: An empirical study. *Total Quality Management & Business Excellence, 21*(9), 953–969.

Meier, H. W., Wyatt, D. J. (2008). Construction Specifications: Principles and Applications: Thomson Delmar Learning.

Mohammed, D. F., & Tan, H. F. (2001). Developing world class construction companies in Singapore. Construction Management & Economics, *19*(6), 591–599.

Oberlender, G., Peurifoy, R. (2010). Formwork for concrete structures: McGraw-Hill Companies, Incorporated.

Peurifoy, R., Schexnayder, C. J., Shapira, A., Schmitt, R. (2010). Construction planning, equipment, and methods: McGraw-Hill Companies, Incorporated.

Tattersall, G. H. (1990). Workability and quality control of concrete: Taylor & Francis, New York.

Chapter 3
Quality Management

3.1 Evolution of Quality Management

The success of the Japanese mass production era in managing quality has led the West, particularly the United States and the United Kingdom, to wake up to the importance of quality (Boje and Dennehy 2008). In 1979, a national uniform standards for quality system, BS 5750, was published in United Kingdom. Soon, the British Standards Institution (BSI) submitted a proposal to the ISO that a new technical committee be formed to prepare international standards related to quality assurance techniques and practices. As a result, the ISO 9000 series was published for use in 1987 and was established in Singapore in 1991. The ISO 9000 series defines comprehensive quality management concepts and guidance, what elements quality systems should encompass, but does not specify how to accomplish them (Gould and Joyce 2003). Hence, a guideline on the implementation of total quality management (TQM) was published by BSI in 1992 and referred to as BS 7850.

TQM is a company-wide effort that involves everyone in the organisation to improve performance (CII and TQM Task Force 1994). It is a management philosophy that effectively determines the customer's needs and provides the framework, environment and culture for meeting those needs at the lowest possible cost. However, Samson and Terziovski (1999) found that construction organisations have not progressed to implement continuous improvement initiatives, and therefore, the potential for learning has been inhibited. Therefore, it suggests that if the construction industry is to improve its quality performance, contractors must learn from their mistakes and adapt to the changing environment by analysing the integration of the company's strategy and structure, technical capabilities as well as quality culture, which is the backbone of TQM (Oswald and Burati 1992). Strategy and structure will address the "what" and "how" contractor manages quality (Jaafari 1997). The technical component will deal with the contractor's skills, practices, tools and methods used throughout the CONQUAS management process (Jaafari 1997). Lastly, having a quality culture will address the norms and behaviour expectations the contractor has set for itself and its people (Jaafari 1997).

S. P. Low and J. Ong, *Project Quality Management*, DOI: 10.1007/978-981-287-074-2_3,

On the whole, TQM has been widely recognised and successfully implemented in many large enterprises, giving them the edge in international as well as local competitiveness through the production of high-quality products to satisfy customer needs (Dale and Plunkett 1990; Hakes 1991; Carr et al. 1996). It is deemed that companies who make use of the TQM concept as their QMS, which is independently assessed for the ISO 9001 requirements, will be recognised for its ability to provide its business outputs to known and consistent standards for performance and quality (Palaneeswaran et al. 2006). TQM is a fundamental component that a construction organisation should embrace if it is to attain excellence in their work performance (Samson and Terziovski 1999). TQM is a way of thinking about goals, organisations, processes and people to ensure that the right things are done right the first time (Low and Teo 2004). Thus, by using TQM as a management concept and CONQUAS as a measurement tool, an efficient and effective QMS can be developed and implemented so as to achieve good construction workmanship quality.

3.2 Rationale for Quality Management in the Construction Industry

The American Society of Civil Engineers (2000) defined quality in construction as meeting established requirements as follows:

> Quality in constructed project is achieved if the completed project conforms to the stated requirements of the principal participants (owner, design professionals, contractors) while conforming to applicable codes, safety requirements and regulations.

Further, it is reckoned that a successful contractor is one who recognises the importance of quality to its activities, understands the need for the proactive management of quality and puts in place the mechanisms to ensure that quality management is undertaken systematically, rigorously and continuously. This indicates that the performance of the contractor and the quality of the building are the most distinguished differentiating characteristics in the construction industry (Ng 2005).

With respect to raising the quality of Singapore's built environment, contractors are constantly challenged to improve their workmanship quality, and hence, they have to be driven to find better ways of undertaking the quality management process. Moreover, CONQUAS, as a well-recognised scheme, will demand contractors to meet the exacting needs and ever-increasing expectations of the quality of construction which are defined and described in the contract specifications. At the same time, BCA has to play a critical role in reshaping the CONQUAS workmanship standards constantly as well as providing other support initiatives such as talks, guidelines and courses in order to help contractors deliver a project within the specified level of quality. With that, quality can then be managed controllably and in an assuring manner so as to achieve high CONQUAS scores and place contractors in the appropriate and fruitful position to serve the construction industry and prosper (Gould and Joyce 2003).

3.3 Quality Control and Quality Assurance

Quality control (QC) primarily deals with issues relating to conformance to the plans and specifications. This means that all of the materials, systems and workmanship applied to the project must be designed to conform to the requirements set forth in the contract documents (Thorpe and Sumner 2004). QC can be accomplished using a number of different mechanisms: submittals, mock-ups, shop drawings, inspections and testing, which are all called for in the project manual (Sukhija 2009). These are the operational techniques and daily activities that can be used to fulfil requirements to achieve CONQUAS standards. Before that, it is essential that contractors promote a quality culture as the support and commitment of the top management is a key factor to make the QC application successful and meaningful (Kanji and Wong 1998).

On the other hand, quality assurance (QA) is a process which verifies that QC has been addressed by dealing with policies and procedures associated with hiring, training, subcontracting and procurement to effectively realise good craftsmanship and workmanship (Thorpe et al. 1996). These must be planned systematically so that necessary actions can be taken to provide adequate confidence that the building will satisfy the given requirements for CONQUAS. Moreover, QA promotes the "right-first-time" philosophy by establishing procedures and defining processes so that checks can be made at key points in the process to ensure compliance with the procedure (Chung 1999).

Hence, it is important that CONQUAS utilises both a project perspective (QC) and a process perspective (QA) in their CONQUAS QMS so as to aid in the achievement of a good CONQUAS score (Jackson 2010). Besides that, having an explicit CONQUAS QMS will certainly bring about huge cost benefits which will be discussed in the next section.

3.4 Pareto Efficiency of Quality Management

An average contractor is estimated to spend 5–10 % of the project cost doing things wrong and rectifying them as they often do not plan work properly the first time and ensure that the required workmanship standard can be achieved in order to avoid the price of non-conformance (Ong 1997). Ong (1997) also mentioned a "Ten Time" rule, which means that the cost of putting right quality problems at the construction phase is ten times higher as it does in the design phase and it costs ten times as much again to wait to resolve these quality problems once the product is in the commissioning phase as compared to if it is put right in the construction phase. This cost includes all cumulative cost factors such as lost man-hours and delays resulting from non-conformance activities such as rework, back charges and rescheduling to expedite construction.

In contrast, the price of conformance includes "doing it right the first time", self-checking, creating procedures and training (Sullivan 2010). The rewards include increased customer satisfaction and the likelihood of repeat business; greater access to public sector and private sector contracts; business processes and procedures become uniform, standardised and consistent; operations can become streamlined and more efficient; management and workforce better understand the objectives of the organisation and their contributions to it; and the reputation and standing of the organisation are enhanced through QMS certification (Griffith 2011). It can be seen that the benefits are many, diverse and powerful.

Theoretically as well as practically, the cost of the conformance is less than the saving in abortive costs of non-conformance (Rezaei et al. 2011). Hence, a substantial time and cost-saving can be realised if the contractor puts in adequate investment to plan and control quality. The cost of initiating a proper QMS is in the range of 0.1–0.5 % of the total project cost, and this has to be monitored closely so that it is within the overall construction and company budget (Sullivan 2010). From Sullivan's (2010) experience, this will trigger savings of at least 0.5–3 % of project cost, a return in excess of five times the investment for both the developer and the contractor. This is because high-quality work results in elimination of rejected work and less intense scrutiny by the architect and engineer, which may have the effect of reducing the client's inspection costs. Furthermore, a high-quality structure will result in reduced maintenance costs over the life of the building. Therefore, both time and cost-savings can improve with proper quality management in the longer term.

In fact, CONQUAS, as a tool of the QMS, can be used to balance between cost, schedule and quality. This is because achieving CONQUAS standards will mean that the price of non-conformance can be avoided as far as possible. The higher score the project garners, the better the quality which also means that construction will progress in accordance with the original schedule and lower costs are expected (Koehn and Regmi 1990). But this deduction is accurate only to a certain extent as after an optimal level is reached, there will be diminishing returns (BCA Academy 2012).

Further, apart from time and cost, quality is also a major dimension of control to help in the process of deducing what the project performance formerly was and what it is now (Fewings 2005). In particular, CONQUAS tells whether and how well the quality performance of the project is met. It helps ferret out the source of problems and design corrective action schemes. Most importantly, CONQUAS measurement highlights where the opportunities for workmanship quality improvement lie so that optimal achievement of high CONQUAS scores can be realised.

3.5 Barriers to Quality Management in Construction

Even with the obvious benefits of quality improvement, quality performance in construction is lagging behind many industries, including manufacturing (Gould and Joyce 2003). This suggests that one of the contributory factors is due to the numerous obstacles that contractors faced in trying to execute quality management practices to attain higher CONQUAS scores.

Hoonakker and Loushine (2010) found that construction projects are exposed to adverse climatic conditions which interrupt the smooth flow of construction works and disrupt its quality management process. In addition, additions or omissions to the original scope of works are common and this will affect the subsequent workmanship quality in the reconstructed area (Robert and Linda 2000). As profit-driven contractors want to minimise cost, they hire insufficient and incompetent staffs to deal with the workload as well as select subcontractor based on the lowest price with no regard to their workmanship quality (Ashford 1989). Similarly, clients do not insist workmanship quality as a top priority and are more concern about the cost involved (Saarinen and Hobel 1990). Further, there is no training and proper directions given to staffs (Kanji and Wong 1998), which means that contractors are unwilling to support the QMS and adopt the tactic to ensure better workmanship quality in order to achieve high CONQUAS scores.

Based on these commonly cited barriers to quality management, it is gathered that in order to maintain control over the quality levels of all the works, there is a need to develop a set of CSFs to overcome these barriers and reach the targeted level of CONQUAS score.

3.6 Application of Quality Management to CONQUAS

In fact, there are many different ways to apply quality management as well as factors to be considered in the application process as investigated by the following quality gurus in Table 3.1. As it is seen, these quality management considerations are very similar to the preceding or following instruments that were developed by other quality gurus too.

Generally, most of the literature concludes that it will be a challenging process to transpose and translate the principles, practices and techniques of quality management to construction (Formoso and Revelo 1999; Lahndt 1999; McCabe 1996; Soares and Anderson 1997). This further reinforced the fact that achieving high CONQUAS score is a difficult process together with the evidence of an extremely low percentage of projects achieving high CONQUAS score as highlighted earlier. Although Singapore's attempt to do so has met with several obstacles, the same conclusion can be drawn in Hong Kong where the performance assessment scoring scheme (PASS), which was developed based on CONQUAS, did not result in a noticeable improvement in the general level of quality even after seven years (Tam et al. 2000). Therefore, drawing on the above studies, the next chapter will identify a set of CSFs, tailored for CONQUAS management by main contractors in the Singapore context, which can be applied to manage CONQUAS effectively and efficiently.

Table 3.1 Quality management considerations

Quality gurus	Quality management considerations	Years
Deming	• Strong management commitment to quality • Process design and control through statistical tools • Continuous search for and correction of quality problems • Purchasing policy emphasising quality over costs • Removal of all barriers to employee participation and teamwork • Effective communication • Eliminating numerical goals and quotas • Company-wide training and quality education	1982, 1986
Saraph, Benson and Schroder	• The role of divisional top management and quality policy • The role of the quality department • Training • Product design • Supplier quality management • Operating procedures • Quality data and reporting • Employee relations	1989
Flynn, Schroeder and Sakakibara	• Top management support • Quality information • Process management • Product design • Workforce management • Supplier involvement • Customer involvement	1994
Black and Porter	• Corporate quality culture • Strategic quality management • Quality improvement measurement systems • People and customer management • Operational quality planning • External interface management • Supplier partnerships • Teamwork structures • Customer satisfaction orientation • Communication of improvement information	1996
Jabnoun and Sedrani	• Customer focus and continuous improvement • Management commitment to quality • Training and empowerment • Benchmarking	2005
Demirbag, Tatoglu, Tekinkus and Zaim	• Quality data and reporting • The role of management • Employee relations • Supplier quality management; training • Quality policy • Process management	2006

References

American Society of Civil Engineers. (2000). *Quality in the Constructed Project: A Guide for Owners, Designers, and Constructors*. ASCE Publications.

Ashford, J. L. (1989). *The management of quality in construction*. London: E. & F.N. Spon.

Boje, D. M., & Dennehy, R. F. (2008). *Managing in the postmodern world: America's revolution against exploitation* (2nd ed.). Information Age Pub Incorporated.

Building and Construction Authority Academy. (2012). Quality and Productivity Seminar, Retrieved November 1, 2012, from http://www.bcaa.edu.sg/QPS2012.aspx

Carr, L. P., Dambolena, I., Kopp, R. J., Martin, J., Rafii, F., & Schlesinger, P. F. (1996). *Total Quality Management: A cross functional perspective*. New York: Wiley.

Chung, H. W. (1999). *Understanding quality assurance in construction: a practical guide to ISO 9000 for contractors*. London: Taylor & Francis.

Construction Industry Institute, & TQM Task Force. (1994). *Implementing TQM in engineering and construction*. University of Texas at Austin.

Dale, B. G., & Plunkett, J. J. (Eds.). (1990). *Managing quality*. New York: Philip Allan.

Fewings, P. (2005). *Construction project management: An integrated approach*. London: Taylor & Francis.

Formoso, C. T., & Revelo, V. H. (1999). Improving the materials supply system in small-sized building firms. *Automation in construction, 8*(6), 663–670.

Gould, F. E., & Joyce, N. E. (2003). *Construction Project Management*. Prentice Hall.

Griffith, A. (2011). *Integrated management systems for construction: Quality, environment and safety*. Trans-Atlantic Publications.

Hakes, C. (1991). *Total Quality Management: The Key to Business Improvement*. London: Springer.

Hoonakker, P., Carayon, P., & Loushine, T. (2010). Barriers and benefits of quality management in the construction industry: an empirical study. *Total Quality Management & Business Excellence, 21*(9), 953–969.

Jaafari, A. (1997). *Proceedings of the International Conference on Leadership and Total Quality Management in Construction and Building*, 6–8 Oct 1997. Singapore: CI-Premier PTE Limited.

Jackson, B. J. (2010). *Construction management JumpStart: the best first step toward a career in construction management*. New York: Wiley.

Kanji, G. K., & Wong, A. (1998). Quality culture in the construction industry. *Total Quality Management, 9*(4–5), 133–140.

Koehn, E., & Regmi, D. (1990). Quality in constructed projects: International firms and developing countries. *Journal of Professional Issues in Engineering, 116*(4), 388–396. doi:10.1061/(ASCE)1052-3928(1990)116:4(388).

Lahndt, L. (1999). TQM tools for the construction industry. *EMJ Engineering Management Journal, 11*(2), 23–27.

Low, S. P., & Teo, J. (2004). Implementing total quality management in construction firms. *Journal of Management in Engineering, 20*(1), 8–15. doi:10.1061/(ASCE)0742-597X(2004)20:1(8).

McCabe, S. (1996). Creating excellence in construction companies: UK contractors' experiences. TQM Magazine, *8*(6), 14-19.

Ng, S. T. (2005). Performance of engineering consultants in ISO 9000-based quality management systems implementation. *Engineering Construction and Architectural Management, 12*(6), 519–532.

Ong, B. H. (1997). *Improving quality performance in construction using a total quality management model*. M. Eng., University of Malaya.

Oswald, T. H., & Burati, J. L. (1992). *Guidelines for implementing total quality management in the engineering and construction industry*. Bureau of Engineering Research, University of Texas at Austin.

Palaneeswaran, E., Ng, T., & Kumaraswamy, M. (2006). Client satisfaction and quality management systems in contractor organizations. *Building and Environment, 41*(11), 1557–1570. doi:10.1016/j.buildenv.2005.06.004.

Rezaei, A. R., Çelik, T., & Baalousha, Y. (2011). Performance measurement in a quality management system. *Scientia Iranica, 18*(3), 742–752. doi:10.1016/j.scient.2011.05.021.

Robert, A. O., & Linda, L. B. (2000). An integrated view of project and quality management for project-based organizations. *International Journal Quality and Reliability Management, 17*(4), 351–363.

Saarinen, A., & Hobel, M. (1990). Setting and meeting requirements for quality. *Journal of Management in Engineering, 6*(2), 177–185. doi:10.1061/(ASCE)9742-597X(1990)6:2(177).

Samson, D., & Terziovski, M. (1999). The relationship between total quality management practices and operational performance. *Journal of Operations Management, 17*(4), 393–409.

Soares, J., & Anderson, S. (1997). Modeling process management in construction. *Journal of Management in Engineering, 13*(5), 45–53.

Sukhija, R. (2009). *Quality management: an excellence model*. Coronet Books Incorporated.

Sullivan, K. (2010). Quality management programs in the construction industry: best value compared with other methodologies. *Journal of Management in Engineering, 27*(4), 210–219.

Tam, C. M., Deng, Z. M., Zeng, S. X., & Ho, C. S. (2000). Quest for continuous quality improvement for public housing construction in Hong Kong. *Construction Management & Economics, 18*(4), 437–446.

Thorpe, B., & Sumner, P. (2004). *Quality management in construction*. Gower: Gower Publishing Ltd.

Thorpe, B., Sumner, P., & Duncan, J. M. (1996). *Quality assurance in construction*: Gower: Gower Publishing Ltd.

Chapter 4
CONQUAS Critical Success Factors

4.1 Introduction

There are a large number of factors determining the workmanship quality of contractors who are expected to perform better and better in terms of their CONQUAS scores. However, contractors can only manage a certain number of factors simultaneously. The vast amount of factors needs to be reduced to some manageable few but critical ones before proper measures can be taken to enhance the CONQUAS score. In addition, the success factors are in the nature of competing limited resources—money, manpower, time and management efforts; investing more on some factors often means less on others. It is thus necessary to identify the vital factors and allocate sufficient resources to those which are the most likely to yield a maximal outcome of high CONQUAS score.

Although quality may be appreciated in the ease of use of a building or it can be seen aesthetically and many more (Robert and Linda 2000), for the most part in this research, the viewpoint of contractors will be portrayed to showcase the CSFs specifically for achieving workmanship quality. According to Merriam Webster, "success" is a "desirable or favourable outcome" which is defined to mean meeting the minimum CONQUAS standard in the context of this research. Therefore, the CSFs are actions which best enable the contractors to accomplish the CONQUAS objective throughout the project; some may be more critical in the early stages while others are more crucial in the later stages. This means that the identification of CSFs should start from the instant when the project was awarded as well as throughout its ongoing development and construction stage till the time when all CONQUAS assessments are completed.

Indeed, the factors of quality success have been reviewed by some authors in their respective findings (Koehn and Regmi 1990) but contractors have not been able to allocate and align their limited resources in an effective and efficient way to the achievement of high CONQUAS scores. This may be due to the weak link and lack of awareness between the CSFs and achievement of high CONQUAS score, and hence, their connection will be described convincingly to determine the

S. P. Low and J. Ong, *Project Quality Management*, DOI: 10.1007/978-981-287-074-2_4,

remedies that contractors have to apply in order to realise the CONQUAS standard successfully. By examining both the characteristics of the construction industry and the factors investigated in other studies, a set of 33 CSFs contributing to an impact on the workmanship quality of the contractor is consolidated into the following five categories:

- Human resource management (HRM).
- Subcontract management (SCM).
- Schedule management (SM).
- Material management (MM).
- Construction management (CM).

4.2 Human Resource Management

The construction sector is one of the most complex and problematic arenas within which to manage people. As a result, the applicability of much mainstream HRM theory to this industry is limited. Indeed, the operational realities faced by construction organisations meant that all too often, the needs of employees are subjugated by performance concerns (Dainty and Loosemore 2012). This has potentially dire consequences for those who work in the construction industry, for the firms that employ them and ultimately, for the prosperity and productivity of the industry as a whole.

In the pursuit to achieve excellent workmanship quality, an organisational structure which involves quality management is one of the first steps that contractors have to adopt, according to the European Foundation for Quality Management framework which also serves as the basis of preparation for establishing the ISO 9001 quality systems (Langford and Rowland 1995). This research has identified seven factors which critically examine key aspects of the HRM function in the context of CONQUAS management and the ways in which contractors respond to the myriad pressures that they face through their HRM practices to succeed in achieving high CONQUAS scores.

4.2.1 Formation of Quality Management Unit with Clear Guidelines of Each Member's Roles and Responsibilities for Every Project: HRM1

Figure 4.1 shows the Quality Management Unit (QMU) structure typically employed by construction companies. It is noted that QMU may be termed differently in different firms but the purpose of this committee and responsibilities of the members remains the same. Due to budget and manpower considerations, studies have shown that none or few contractors have employed full-time site staffs

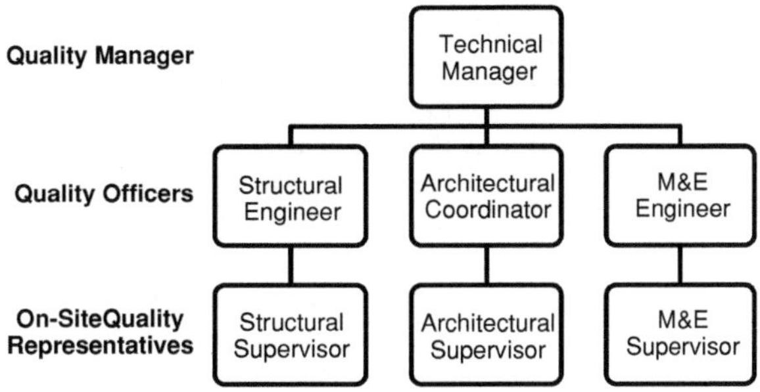

Fig. 4.1 Composition of a typical quality management unit. *Source* CIDB Manual (1989)

just to deal with quality management issues unless required by the client (Rounds and Segner 2010). Hence, team members would have to take on quality management as one of their roles in the project to make sure that construction work is carried out correctly from the very beginning. This is to ensure that good workmanship quality will be adhered and not neglected.

The QMU is vested with the necessary authority to approve and control work procedures and also to report on the quality performance of the project that does not meet quality requirements with the minimum standards being based on CONQUAS (CIDB 1989). There are, therefore, good reasons for contractors to utilise the QMU approach to quality management and cope with the CONQUAS requirements effectively.

The QMU is a team and collaborative effort. Therefore, besides the formation of a QMU, there must be clear roles and responsibilities assigned to each member to avoid any misunderstandings and negligence in accomplishing their respective job scope. Further, this is to prevent the QMU of the contractor's firm from assuming that this is the responsibility of the subcontractors and labourers who tend to disregard the importance of achieving workmanship quality as long as the job is more or less done (Palaneeswaran et al. 2006). Similarly, as long as the project can be completed on time and within budget, clients are likely to overlook on the quality aspect (Rommel 1996). Therefore, the findings suggest that clients should be included as part of the QMU to contribute and supervise the CONQUAS management workflow.

In addition, the roles and responsibilities assigned to the QMU have to be actively pursued and consciously managed (Griffith 2011). This is because having just a few people abiding by the QMU approach is insufficient as achieving the workmanship quality requirement is not a single responsibility of any party but requires a team effort from all stakeholders of the project (Saarinen and Hobel 1990). Therefore, proper management of this factor is essential to discover any non-compliance to workmanship quality and achieve high CONQUAS score. An example of how the samples required can be prepared for CONQUAS submission is shown in Table 4.1, and each of them should be specifically allocated to the respective person in charge to make sure that none of the tasks are neglected. Overall, this shows the importance of clear distribution of work responsibilities to avoid any ambiguity that would impede the progress of the quality work performance.

Table 4.1 Tasks allocation for sample preparation

Component	Tasks	Person-in-charge
Structural works	1. Reinforcement bars, formwork	
	2. Finished concrete	
	3. Precast concrete (if applicable)	
	4. Structural steel assessment (if applicable)	
	5. Pre-stressed concrete (if applicable)	
	6. NDT-covermeter test (CVM)	
	7. NDT-uniform pulse velocity test (UPV)	
	Submit FORM A—cube and tensile steel (no attachment needed)	
	Submit FORM B—welding test declaration [attachment needed—magnetic particle inspection (MPI), non-destructive test (NDT) report]	
Architectural works	1. Internal finishes (principal, service and circulation areas)	
	2. External walls	
	3. External works	
	4. Roof	
	5. Windows' water tightness test—by BCA	
	6. Windows' water tightness test—self-test by contractor	
	7. Wet-area's water tightness test—by BCA	
	8. Wet-area's water tightness test—self-test by contractor	
	9. Pull-off test (POT)	
	10. In-process wet-area water tightness test	
	Submit FORM C—Singapore Concrete Institute (SCI) wavier for in-process assessment (attachment needed)	
	Submit FORM D—Material and functional test (no attachment needed)	
M&E works	1. Electrical (both during structural and architectural assessment)	
	2. ACMV	
	3. Plumbing and sanitary	
	4. Fire protection	
	Submit FORM E—M&E performance test (no attachment needed)	
	Submit FORM F—Draft CONQUAS Certificate	

Source BCA (2012)

4.2.2 Commitment of Top Management to Advance the CONQUAS Management Performance: HRM2

Top management commitment plays a vital role to advance the CONQUAS management performance, and this has been cited by practicing quality gurus such as Deming (1986) and Juran (1986) as one of the most important factors impacting the success potential of TQM in a firm and improving workmanship quality performance (Anderson et al. 1995; Wilson and Collier 2000). Successful implementation of TQM requires effective change in an organisation's culture, and it is almost impossible to change an organisation without a concentrated effort by the top management who aims for continuous improvement, open communication and cooperation throughout the value chain (Daft 1998; Zeitz et al. 1997). Thus, top management should be actively involved with and dedicated to improve quality within the organisation and to subsequently improve the CONQUAS score.

Moreover, Ahire and Shaughnessy (1998) conducted a large-scale survey which found that firms with high top management commitment implement the other TQM implementation elements more rigorously than those with low top management commitment. In the context of this research, this means that different levels of top management commitment will influence the dynamics of implementing the other CSFs which is essential to the achievement of high CONQUAS scores. The role of top management commitment in CONQUAS management efforts is hence pivotal to advancing CONQUAS performance.

In addition, when top management first learns about TQM, they are often quick to "convert" (Ahire and Shaughnessy 1998). However, a top management team which adopts this approach without a serious evaluation of the efforts and resource requirements heads quickly towards failure (Jelinek and Adler 1988; Dutton and Ashford 1993). Top management must go beyond merely adopting slogans of improving quality and actually become involved in quality efforts at various planning, implementation and monitoring phases. The impact top management commitment has on the resultant quality has also showed that in the long run, superior and consistent quality leads to improvements in cost and delivery performance (Ferdows and Demeyer 1990; Leach et al. 1993).

Overall, when the top management is committed, they will put in effort to build a quality culture by creating clear values and beliefs so as to foster total quality behaviour (Linklow 1989). This is the main ingredient in a successful TQM programme (Westbrook 1993) which is recognised by many quality experts such as Deming, Juran and Crosby. Their works also identify a number of cultural elements within the company that must undergo change in order to sustain quality improvement efforts (Sommerville and Sulaiman 1997). Undoubtedly, the success of high CONQUAS score needs the active commitment and involvement of the top management to create a quality culture among the project team which enables them to work together effectively to attain quality goals.

4.2.3 Getting the Support and Cooperation of Employees: HRM3

Although leaders provide direction and resources supporting quality, higher-quality products can only be achieved if they receive the support and cooperativeness of employees. Participation in the field will help in driving the quality management practices throughout the entire company such that each and every employee understands the importance of achieving workmanship quality (Pike et al. 1994). This is because workmanship quality is ultimately achieved at the labour force level (Hernandez and Aspinwall 2008).

Employees and subcontractors alike must embrace the quality management practices and endeavour to produce high-workmanship-quality work. Employee involvement strategies like quality control circles (QCC) and cross-functional teams have been used successfully by many organisations to realise gains in different phases of quality improvement (Ahire and Shaughnessy 1998). For example, cross-functional teams have been used in new product development and distribution management (Carmel 1995). QCC have also been used with mixed success in quality efforts (Dale and Duncalf 1984). It has also been demonstrated in practice that the judicious application of these tools combined with team-based incentives can result in useful suggestions that improve the quality of processes and products (Cole et al. 1993). Further, encouraging and recognising their hard work will motivate them to cultivate a "first time right" approach to take the project forward.

In addition, employees can be empowered to make decisions related to quality by assignment of responsibility for the quality results, and provision of technical and managerial support to aid workers in quality efforts (Ahire and Shaughnessy 1998). It has been argued that these elements lead to more and better worker involvement in quality improvement efforts (Harber et al. 1991) in world-class organisations like Toyota (Everett and Sohal 1991). Therefore, it is evident that the attitude towards quality must not only emanate from the highest levels of leadership but also from the middle management down to every single worker and subcontractor in the field, in order to realise the goal of attaining high CONQUAS scores.

4.2.4 Training Employees to Understand and Adhere to CONQUAS Requirements: HRM4

Although CONQUAS dictates the expectations for quality and "what is right", studies have shown that many contractors do not understand what it actually means (Fouayzi et al. 2006). This reinforces the need to train and equip employees with the information and skills needed to enforce and ensure that workmanship quality is good right from the first time. Bell and Burnham (1989) also support this notion and stated that it is impossible to improve any organisation's operations without a well-trained workforce. Training is defined as the process of developing work-related knowledge and skills in employees for the purpose of improving performance systematically (Saraph et al. 1989). Effective training methods include simulation exercises, and case studies which can be used in the construction industry to create learning situations based on experience (Tabassi and Bakar 2009).

In fact, BCA has organised CONQUAS training workshops for contractors as well as developers and consultants. The focus area of the contractors training is to educate them on the adequate skills required based on the assessment standard and make known to them on areas that BCA assessors look out for which are also common areas with workmanship quality problems. This will allow contractors to conduct an in-house assessment to simulate the result and rectify non-compliances before the actual assessment to improve their CONQUAS performance. On the other hand, the focus area of the developers and consultants training is to teach them to improve construction quality by adopting the right CONQUAS strategy upstream and help contractors to produce quality buildings. This is important as consultants are involved in the drafting of contract specifications and plans, and hence, they have to understand the minimum requirements of the workmanship quality standards.

By having an understanding and familiarisation with the CONQUAS standards, the appropriate resources to achieve the stipulated workmanship quality can then be procured. Hence, the top management has to provide the resources necessary for training employees in the use of new quality-related principles and tools and to create a work environment conducive to employee involvement in the process of change. Furthermore, it is especially significant for employees and, in particular, the QMU to know what represents high or low quality when it comes to workmanship as they are the ones who will be performing quality checks and determine whether they are ready for BCA assessment.

4.2.5 Supervision by Client or Architect to Monitor the CONQUAS Management Workflow of the Main Contractor: HRM5

Ideally, the employers (client or client's representative such as the architect) should be actively involved to make sure that the contractor's quality plan will work in the best interest of the project's quality targets (Brennan 2008). It is important that they provide the downstream construction team with a clear quality performance standard and require them to attain a benchmark higher than the norm. This will push them to do their utmost effort in executing quality management practices which are essential to the attainment of high CONQUAS scores. Also, they should render any assistance to the contractors and be open to suggestions and changes due to design flaws to help achieve the CONQUAS standards. Contrary to popular belief, this shows that the accountability for achieving high CONQUAS scores has to be shared with the client as well even though the contractual responsibility still lies with the contractors.

However, many have ignored the upstream involvement of clients and architects to also put in force a quality culture in the project besides the part on the contractors (Kanji and Wong 1998). This is because of the fact that employers will be in a contradicting position to convey the philosophy that quality matters. By assigning a higher priority to quality over cost or schedule (Ferdows and Demeyer 1990) and providing adequate resources to the implementation of quality management efforts, requires them to invest in extra human and financial resources (Chapman et al. 1991) and hence, they are unlikely to do so. Thus, the findings indicate that much government effort is needed so as to motivate developers and architects to take up this task of monitoring the CONQUAS management workflow of the main contractor. This is because when contractors are under stringent supervision by another party, developers will benefit from the deliverance of a building with a high CONQUAS score. Hence, they certainly have a role to play in the achievement of high CONQUAS score.

4.2.6 Comparing Company-Wide CONQUAS Performance Track Records: HRM6

In addition, contractors can learn from the strengths and weaknesses of their past projects to serve as inputs for their current and future projects. With that in mind, the possibility of repeating the same defects and non-compliance can be reduced. On a corporate level, the quality performance of all projects should also be reviewed so that contractors can reflect on and exchange pointers with each other on how to score better in future projects. However, concrete empirical studies have not been tested to prove that the more skilled and knowledgeable the individual is, or the more experienced the company is, the better it is at striving for construction quality excellence. This is seen as even the top performers have varying CONQUAS scores (Corenet 2012) which may be due to the fact that the projects are completed by different teams of people.

4.2.7 Setting a Benchmark to Gauge and Control the CONQUAS Performance: HRM7

According to Camp (1989), benchmarking is the search for industry best practices that will lead to superior performance. This proposes that contractors identify their key business processes and measure and compare them against other contractors considered to be world-class performers. It is noted that by improving and developing key processes in this way can provide a route to gauge and control the CONQUAS performance. Therefore, in order to drive and reinforce the importance of achieving high CONQUAS scores, the company's CONQUAS management policy should set high CONQUAS targets. To do so, contractors can develop their own benchmarking system or rely on the online IQUAS benchmarking tool as a platform to scale their CONQUAS performance as well as take note of major trends on workmanship quality and industry good practices to address common defect areas effectively for continual improvement. To put it briefly, Fong et al. (1998) have categorised the benchmarking process into five phases as illustrated in Fig. 4.2.

Moreover, benchmarking can be used for achieving awards such as the BCA CEA and Green and Gracious Builder Award or against international standards

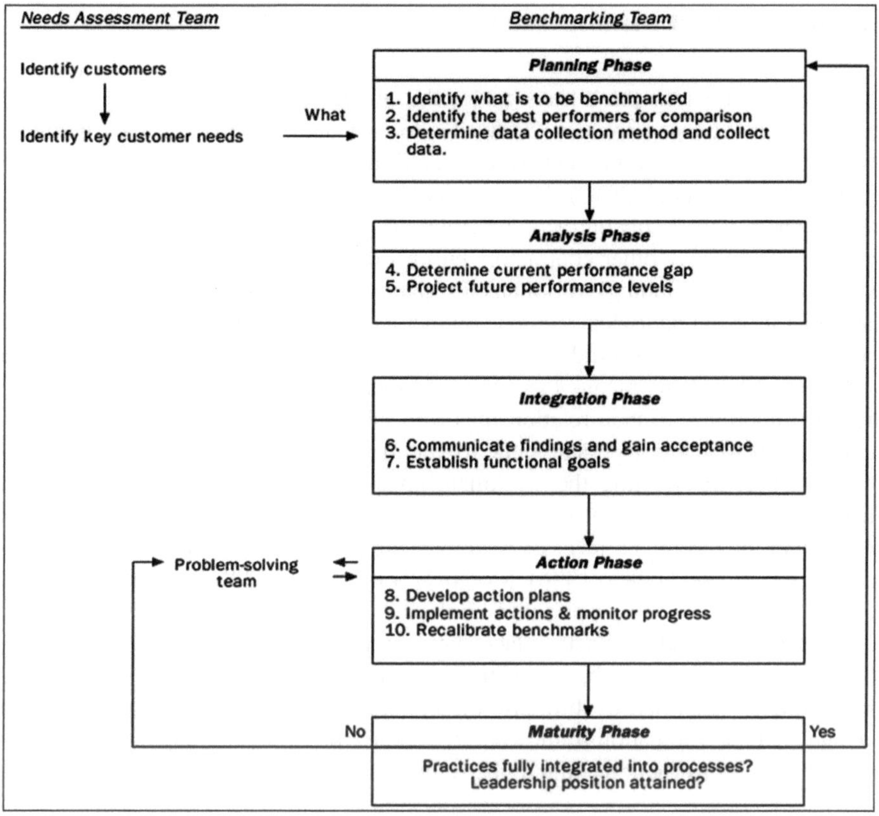

Fig. 4.2 Benchmarking process model. *Source* Fong et al. (1998)

or, in their absence, evaluated against internal standards in the contractor's firms (Wilson and Durant 1994). Past project performance by other contractors can also share their strategies that have allowed them to achieve high CONQUAS scores. Notably, Woh Hup Pte Ltd who has received 34 CEA seeks to meet the changing demands of the construction industry and consistently deliver high-quality building solutions. Their staffs are trained to ensure that projects meet stringent quality control guidelines and overcome construction difficulties through innovations to the construction process and good project management. Additionally, consistency in workmanship quality and continuous improvement of a QMS can be achieved by benchmarking the GIP to check if those practices are really implemented or not.

Indeed, benchmarking can be seen as an important management tool of TQM (Kouzmin et al. 1999), to measure the various elements that play a part in affecting the CONQUAS performance and determine an appropriate quality target to be set. All that said, it is important that accurate benchmarking along with feedback and continuous learning needs to become a genuinely adopted practice in order to achieve a high CONQUAS score that is attainable within the resource constraints faced by the contractor.

4.3 Subcontract Management

Few main contractors these days undertake all the work involved in a contract on their own account (Brennan 2008). To do so would require that they maintain resources of men and equipment which would inevitably be under-utilised much of the time (Peurifoy et al. 2010). Thus, subject to the provisions of the main contract, they prefer to award subcontracts, whereby all or particular elements of the work will be performed by other contractors but this does not relieve any of his contractual obligations, including workmanship quality. Further, Matthews et al. (1996) stated that it would be logical if main contractors want to improve their performance and productivity, they should concentrate their efforts where the majority of the work takes place. This means that main contractors should increase the depth and strategic importance of their relationships with subcontractors too (Eom et al. 2008).

Subcontractors are like suppliers to the main contractors and both play an important part in affecting the workmanship quality of the construction process. Therefore, it is essential that subcontractors also have their own quality programmes to ascertain an acceptable level of workmanship quality of the final building product (Ghobadian and Gallear 1996). Moreover, Eom et al. (2008) found that the main contractor places primary importance on the subcontractor service and financial stability, whereas the subcontractor places primary importance on technical capability, competitiveness, self-growth, financial growth and site process. With that, nine factors have been identified to showcase the importance of developing a subcontractor management and evaluation framework to enhance the overall quality and productivity level of the construction chain to the achievement of high CONQUAS scores.

4.3.1 Controlling the Number of Subcontractors Working on the Project: SCM1

Having numerous subcontractors on each project is common which also makes it more difficult to control the workmanship quality of each subcontractor (Fewings 2005). This is because each subcontractor is only concern with their respective scope of works and tends to ignore the effects on the subsequent works of other contractors. If the first trade was not properly completed, the workmanship quality of the subsequent trade will be affected. This can create major problems in the achievement of the final workmanship quality unless an adequate pre-contract selection of the subcontractor is established.

Hence, it is reckoned that maintaining a small number of subcontractors will improve the built quality and productivity of the project (Ansari and Modarress 1990). Additionally, dealing with a small number of subcontractors facilitates the solution of quality and delivery problems because main contractors can pay close attention to each subcontractor (Burt 1989). Consequently, the main contractor would then be able to monitor its CONQUAS management workflow better and more thoroughly to ensure that the excellent construction workmanship quality can be realised.

4.3.2 Having a Rigorous Prequalification Process to Select Subcontractors and Suppliers: SCM2

Besides settling on fewer subcontractors and suppliers who can deliver increasingly better quality, they should also be screened thoroughly (Barrier 1992). In fact, during the tendering stage, main contractors should have sought quotations from their approved list of subcontractors, and hence, award the appropriate subcontracts upon award of the main contract. This list of approved subcontractors is constantly updated to integrate more partners to the company as they have demonstrated both the means and the will to meet a high-quality standard during past projects or due to referral from reliable sources. These subcontractors and suppliers who have helped to attain good workmanship quality should then be documented for future reference while those with poor performance should be avoided in future projects. Alternatively, new subcontractors can submit their quality qualifications for approval, especially their methods and measures which have been successful in garnering a high CONQUAS score in their past projects. Other qualifications include their financial stability, details of their company training programmes as well as manufacturer or trade association endorsement to document their capability for executing good workmanship quality results.

Levy (2009) believes that evaluating the subcontractor assigned to the project is as important as the initial contractor evaluation process. Apart from meeting the targeted CONQUAS score, subcontractors that enjoy good reputations for high quality usually go way beyond the mandated methods spelled out in the contract documents. They implement processes that also help them achieve a level of workmanship quality beyond the minimum, resulting in less rework and ultimately more repeat business with the main contractor and client (Rommel 1996). They will also have adequate labour, equipment and other necessary resources to provide construction-related services of the capacity and quality required for compliance with the CONQUAS standards.

Therefore, it is best if the subcontractor and supplier selection process prioritises quality and delivery performance over price when selecting subcontractors and certifying suppliers for material quality (Trent and Monczka 1999). However, there are situations where clients nominate subcontractors themselves, usually in the case of specialist works or due to their lower tender sum in that trade, and hence, the main contractor just have to award the subcontracts to them regardless of their workmanship standards and yet still be responsible for the quality of their work.

4.3.3 Ensuring the Skill Level of Labourers: SCM3

Besides having ISO certification which is a basic criterion as viewed by Din et al. (2011), the subcontractors should also have a good record of employing capable and trained labourers to assure good workmanship performance. This can be achieved by having active partnerships with regular suppliers and subcontractors which will guarantee to a certain extent that the skilled labourers and quality materials are employed to carry out the construction works (Fewings 2005).

4.3.4 *Collaborative Efforts Between Subcontractors and Main Contractor: SCM4*

As important as it is to investigate the subcontracting selection process and ensure the skill level of labourers, it is also necessary to improve methods for enhancing collaborative efforts between subcontractors and main contractor (Eom et al. 2008). Collaborative efforts can be built from a long-term working relationships, and this will ensure that there is adequate commitment and attention received from the subcontractors to achieve the desired workmanship quality in the respective scope of works. A number of researchers have found that improved subcontractor relations enhance the performance of both subcontractors and main contractor, and this is especially true when quality delivery is their priority (Shin et al. 2000); Fewings (2005) also reiterated that it is much safer to work with a repeated company than a new company as it has been known that their businesses are stable and workmanship quality is dependable, respectable and reliable. It can also be ensured that suppliers have adequate inventory, service personnel and distribution resources to deliver the specified products in adequate quantities to service the project according to the contract documents and the construction schedule. Therefore, managing supplier relationships strategically is essential to the success of quality performance because having such a partnership will necessitate a high level of commitment and obligation to perform well so as to create opportunities for future collaboration (Ellram 1991).

Unfortunately, relationships between main contractors and subcontractors are often strained and adversarial (Dainty et al. 2001). Increasing complexity, the oversupply of specialist firms and declining construction output have all contributed to the current antagonistic atmosphere (Kumaraswamy and Matthews 2000). Thus, many studies have suggested adopting a partnering philosophy to overcome these difficulties including elements such as commitment, equity, common goals, communication, trust, cooperation and continuous evaluation (Ogunlana 1991; Ho et al. 2000). This is because not only can such collaboration greatly improve CONQUAS performance, it can also directly benefit the entire construction chain (Wood and Ellis 2005). Therefore, the subcontractor evaluation and management processes must include factors that will enhance cooperative relationships, in particular, sharing mutual objectives, improving communication, participating in collaborative work and developing cooperative relationships (Eom et al. 2008).

4.3.5 *Adequacy of Contract Period and Contract Sum: SCM5*

It should also be noted that the period of construction set has to be adequate as even reputable subcontractors will definitely have to compromise on the workmanship quality in order to meet an extremely tight construction schedule. This is to avoid having to pay for liquidated damages, which is a significant concern of the subcontractor firms who are usually financially constrained (Mincks and Johnston 2004). Similarly, the contract sum set for the project has to be realistic according to the requirements as an acceptable level of workmanship quality can only be met with an adequate contract sum awarded. This is to ensure that there will be sufficient funds available to purchase quality materials as well as employ a skilled workforce to deal with the CONQUAS management workflow.

4.3.6 Identifying a Specific CONQUAS Score that the Subcontractor is Liable for Achieving: SCM6

Thereafter, it is critical to impose workmanship quality standards as part of the terms and conditions in the subcontracts to hold the subcontractors responsible for meeting the CONQUAS requirements as well. The purpose of this is to hold each subcontractor responsible for their respective scope of works and also safeguard the interests of the subsequent subcontractors to ensure that their workmanship should not be affected due to the poor quality performance by the previous trades. This can be implemented monetarily in the two following factors.

4.3.7 Awarding Incentives to Subcontractors if Their Targeted Score is Met: SCM7

An incentive can be awarded if the targeted CONQUAS score is met. However, the initial contract sum of this approach is usually lower so that subcontractors will be enticed to earn the extra incentive by ensuring superior workmanship quality (Levy 2009).

4.3.8 Imposing a Penalty to Subcontractors if Their Targeted Score is not Achieved: SCM8

Another approach will be to award a higher contract sum to help subcontractors make up for the extra cost required in ensuring workmanship quality but there will be disincentives if the targeted CONQUAS score is not met which will push them to work towards attaining good workmanship quality in order not to be penalised (Levy 2009).

4.3.9 Giving Clear Instructions to Subcontractors on How to Adhere to the CONQUAS Requirements: SCM9

A research by Abu-Taiseh et al. (2009) identified that to improve quality and productivity, the main contractor should provide clear instructions to the subcontractors on how to adhere to the CONQUAS requirements in their work processes, and this should be specified contractually. This also calls for the need for main contractors to communicate with the subcontractors frequently, to notify them to pay attention to areas prone to defects non-compliance and if necessary, allocate more manpower to step up on the inspection frequency. This ensures that the subcontractors will do their best to complete the works on time, adhere to all contract terms and conditions and as far as possible, keep to the best quality of work. Communicating a clear strategy for improving quality to the subcontractors can also be enhanced by instituting quality-based incentives and compensation procedures as mentioned. Such efforts to improve will result in mutual growth, benefiting both the main contractor and subcontractor (Eom et al. 2008), consequently, leading to the achievement of a high CONQUAS score.

4.4 Schedule Management

The best schedule is not the schedule showing the project completed in the shortest time period. It is the schedule that is able to meet the quality requirements of the project based on the client's expectations. However, it should be noted that quality expectations must be realistic if time and cost objectives must also be met. Schedules also serve as a communication tool among project participants to allow them to identify potential problems early as well as coordinate various activities efficiently. By establishing the start, duration and completion date of each activity, all stakeholders will know the schedule and whether the work of a particular job can be accomplished in the context of all other work scheduled. Construction projects involve many players and activities, resulting in the overlapping of works. Hence, coordination is vital in order to ensure that work will be carried out smoothly, and within the stated schedule for CONQUAS assessment to take place. Here, four factors have been identified as critical which will allow quality performance as everyone will then know the priority of who should do what, when and where, ultimately enhancing their workmanship quality.

4.4.1 Prefixing of CONQUAS Sample Locations (Applicable to Structural Components Only): SM1

As soon as the project is awarded, schedules are planned. However, schedulers often disregard the need to schedule the construction process to coordinate with CONQUAS assessment. With the exception of architectural and M&E works which are assessed on a random basis, it is definitely beneficial if sample locations for structural works are carefully planned. As structural works will eventually be covered and cannot be seen on the final building product such as formwork, rebars and finished concrete, CONQUAS assessment has to take place before the next work activity in that component continues. Yet, this should not have any effect on the construction schedule. Therefore, prior arrangement is necessary to schedule the assessment to take place as soon as the respective works are completed and before the next trade of works begin.

Subsequently, the BCA assessors will pick 50 % of these planned structural locations randomly on the day of assessment. This implies that measures can be taken to ensure that the planned structural locations are completely compliant with the workmanship standards before CONQUAS assessment. Measures can be taken to prevent or correct the occurrence of any possible defects to perfect the structural score component. This is especially important as such structural concealed workmanship defects would not be discovered following an inspection of the building, and hence, it is crucial that contractors aim to perfect the score of the structural components so as to prevent latent defects from appearing years later.

This can be achieved by paying extra attention on the locations to be included in the sample size so that the workmanship quality of the structural samples taken can be guaranteed to be of high quality. Another strategy is to choose the sample locations where new formwork will be used which the quality trends have shown that there will be a higher chance of achieving full compliance. The QMU will then have to keep an eye on these locations especially during the process of concrete casting. The QMU will also have to make sure that there is sufficient concrete from the same batch of truck for each structural component to prevent settlement from occurring. This is because the process of waiting for the next batch of concrete to arrive or having slightly different concrete quality from different batches will affect the workmanship quality of the finished concrete (Tattersall 1990).

In contrast, the contractor will have to be extremely rigorous in ensuring the workmanship quality for architectural and M&E works as all locations constitute the sample size, and it is entirely up to the BCA assessors to randomly select the areas to be scored. Nonetheless, it is still important to employ TQM efforts for every elements of the project, working hand in hand with the project team, clients and labourers to put into action the quality culture as well as the execution of GIP to the attainment of an overall high CONQUAS score.

4.4.2 Adequacy of Sample Locations Prepared in the Event of Unforeseen Site Conditions to Serve as a Backup: SM2

Several factors can affect the workmanship quality of the schedule of sample locations planned. There may be change in orders and uncertainties due to discrepancies between the actual site conditions and drawings (Jackson 2010). Unexpected weather conditions will also affect the construction schedule and concerted efforts to ensure the workmanship quality in these locations may become ineffective and have to be eliminated from the sample size to remove any possibility of non-compliance. Therefore, more sample locations have to be prepared as a backup to prevent the postponement of construction schedule from affecting the workmanship quality of the sample locations prepared.

4.4.3 Arranging for Phased Construction to Reap Learning Curve on Workmanship Quality: SM3

Commonly, phased construction is adopted as developers want to take over a certain part of the project first so that pre-sale can begin at the earliest possible time for the completed phase. This will tie the buyers down as well as help to prevent or mitigate any financial risks that the project will incur especially in the initial stage, where there may be negative returns and inclination to failure due to uncertainties. But what has been overlooked is that phased construction can reap the benefits of a learning curve as most doubts would have been solved during the initial phase, and hence, subsequent phases of construction would be able to meet or even surpass the required workmanship quality easily (Bennett 2012).

Phased construction speeds up the learning curve effect on employees, thereby enhancing quality and reducing costs due to decreased variety and increased volume (Tan 2001). The failure rate of having defects occurring also decreases and its reliability increases (Ahire and Dreyfus 2000). To a certain extent, this is a preventive approach to quality improvement by having stable production schedules and work distribution (Saraph et al. 1989) to reduce construction variation, resulting in increased output uniformity as well as reduced rework and waste because quality problems were identified and corrected immediately (Anderson et al. 1994). Therefore, by taking steps to prevent the workmanship oversight made during the early stage of construction, the attainment of high CONQUAS scores for the later phases of construction would be easier.

4.4.4 Booking of Assessment Schedule to Tie in with Site Progress: SM4

The timing of calling for BCA inspections is another critical point. This is because wear and tear cannot be totally avoided and minor deterioration will still be expected even with good workmanship quality (Keeble 2006). Therefore, before any signs of deterioration start to appear, contractors should quickly arrange for CONQUAS assessment to reduce the chances of any decline in workmanship quality which is inevitable as time passes.

4.5 Material Management

Once the materials are selected and approved, it is important to determine the right method and sequence of construction where selection of the appropriate equipment is also required. This is because if any construction steps are missed out or inefficient equipment is chosen, the effect on workmanship quality will be detrimental (Rounds and Segner 2010). As a result, even if high-quality materials and a skilful workforce are employed for the works, a good CONQUAS score will not be expected. Therefore, the contractors and subcontractors either have to seek for the supplier's help to determine the construction method or come out with their own method statement based on their understanding and experience of executing such construction works.

Basically, there are two types of deficiencies which arise as a result of non-compliance to the choice of materials and methods of construction. Material deficiencies result in defects that can occur when substandard materials are utilised in the building. Construction deficiencies result in flaws that occur due to poor quality workmanship during the actual construction process. Here, six material management factors have been recognised as critical to the achievement of high CONQUAS scores.

4.5.1 Choose Materials Through a Comparison Process, Considering Their Specifications and Samples to be Provided as Well: MM1

The contractors are required to select and procure suitable construction materials so that they can meet the contract specifications. Unless a specific brand and model number is stated, a thorough study and analysis of the different material properties as well as the reputation of its provider should be done to check for its compatibility in the different zones of the building (Peurifoy et al. 2010). In this quality check to verify the product's quality performance, the contractor is responsible for getting samples if possible and documenting specific data of all the materials to ensure that they comply with the requirements set forth in the contract documents. Then, they can be submitted for approval before they can start to order the materials. In general, it is much better if at least an alternative set of data and sample which has also been investigated for its high quality by the contractors are also provided in the submittals for the architect or client to compare and verify the quality level of the materials. This will enable them to choose the one that is more appropriate to the achievement of the desired CONQUAS score.

Conscientious contractors and subcontractors will conduct this as part of their QA plan which reinforces the importance of prequalifying subcontractors before award of the subcontract (Thorpe et al. 1996). More often than not, contractors simply seek for the architect's or client's approval and pass on the responsibility of quality confirmation to them with minimal information collected which may be hard for them to do a proper quality evaluation (Thorpe and Sumner 2004). Consequently, the resulting degree of the product's workmanship quality upon construction may be far-fetched from the original submittals and have to be reconstructed in order to attain the targeted CONQUAS score. Hence, it is important that materials are methodically chosen so as to avoid using poor quality materials which have an impact on the final workmanship quality of the built product.

4.5.2 Select Materials Which can be Better Managed During Construction: MM2

In addition, the use of factory-made products where possible such as precast concrete or prefabricated components and less cluttered materials such as pre-pack mortar is encouraged as the quality of the end product can be better controlled under factory production conditions which assures a higher number of compliance with the CONQUAS standards. Furthermore, precast components have better quality surface finish with faster construction speed and that one structural bonus point under the CONQUAS eighth edition will be awarded if precast elements are supplied by an accredited precaster. One architectural bonus point will also be awarded for the use of prefabricated bathroom (at least 65 % of the toilets) as it ensures tile joint consistency and has less lippage or unevenness to make sure that there is consistent and quality workmanship. In addition, up to one architectural point will be awarded for selecting drywall partitions as it is easy to install, allowing neat concealment of services, lightweight and strong; using precast or system formwork or cladding façade as it does not require scaffolding and avoids any wet trades. Next, the use of rectified tiles will ensure better dimensional and tonal consistency, as well as better joint consistency. All these will result in better quality outcome and enhancing productivity too.

Inevitably, there are still many elements which have to be installed by hand and, hence, having a skilful workforce or employing the appropriate technology to do so will also ensure higher workmanship quality and increase productivity. Further, mentioned that material selection and usage issue can be addressed through careful drafting of contract with the subcontractors, wherein raw material specifications were covered in detail along with the sampling plan and method statement. Subcontractors can also be guided to select their suppliers so as to ensure conformance to contract conditions. Lastly, information available in respect of the pool of adequate suppliers can be shared with the subcontractors to assure the workmanship quality standards of the material selected.

4.5.3 Inspect Materials Upon Delivery: MM3

First, it is noted that the production of quality building is necessarily dependent on the timely deliveries of quality materials, so that the materials supplied can meet the owner's specifications and standards for quality (Grieco and Gozzo 1985). Once the materials are delivered in a timely manner, its condition has to be properly checked for any damages which may compromise on the constructed workmanship quality. If the materials are not as specified and approved, they should be rejected on the spot and sent back to the supplier. For example, a moisture content check should be done on the material such that it is within 10–14 % for air-conditioned buildings and 14–15 % for non air-conditioned buildings. Flooring dimensions should also be checked such that the width and thickness are within ±0.75 and ±0.4 mm, respectively. Most importantly, they should be defects free and should be wrapped in original packaging condition with seals and labels intact. Therefore, it is certainly critical to conduct an inspection on the materials once they are delivered to site to check whether the quality standards are complied with.

4.5.4 Proper Materials Handling and Storage: MM4

Proper protection during storage is often ignored, and this is also one of the sources for the lack of conformance to the CONQUAS requirements (BCA 2005). There should not be any transportation, loading and unloading in the rain and materials should be directed to area of installation immediately. Storage area should be enclosed, clean and dry, with good air circulation and for some materials, need to be stacked on pallets, not more than a certain safe height, to prevent dampness or if kiln dried, remove packaging only before installation or if air dried, minimum two weeks acclimatisation on site (BCA 2010). This is to prevent any contamination by storing materials off ground. As such, all these examples showed the importance of proper material handling and storage so as to prevent the workmanship quality of materials from deteriorating and affecting the CONQUAS score.

4.5.5 Protection of Materials After Completion of that Portion of Works: MM5

Similarly, upon completion of certain works, adequate protection of that portion of the works is required before completion of the entire project to prevent the construction of other trades from affecting the workmanship quality of the completed trades. For example, there should be no traffic for a certain number of days after completion of certain portion of works, and only light foot traffic is allowed after that to protect the freshly laid material against stepping. Proper barricade should also be set up during the curing of waterproofing membrane to ensure that there is adequate protection so that the waterproofing membrane will not be damaged, contaminated or disturbed. Such measures serve to ensure that the works of the labourers on site will not cause any defects or deterioration to the components which are already constructed.

4.5.6 Sample Testing of Materials Through an Independent Testing Agency (ITA) to Check for Proper Usage of Materials: MM6

Qualified independent agencies will have to be employed to conduct testing and inspection procedures in the field while work is in progress as required by the CONQUAS standard. Sometimes, samples are collected on-site and taken back to be tested in special laboratories. The products and installations that do not meet the test standards specified must be removed and re-done according to the recommendations of the test report, thus establishing the workmanship quality by actual measurements besides just conducting visible inspections. At the same time, it should be noted that defects rising from destructive testing procedures will have to be repaired. Hence, it is also important to coordinate testing procedures with other construction activities to minimise disruption to the work. In this way, the testing standards serve as a quality measurement tool to help contractors in maintaining their level of workmanship quality.

4.6 Construction Management

Construction management is also a CSF to secure a high CONQUAS score. The bulk of construction happens on site, and hence, site works are effectively the direct factors which account for the quality performance of the end product (Mincks and Johnston 2004). The challenge is to make sure that there is quality management of the entire construction process so that workmanship quality can be maintained. Hence, seven construction management factors have been identified as critical to the achievement of high CONQUAS scores.

4.6.1 Ensuring that Shop Drawings are Checked Thoroughly Before Actual Construction: CM1

Generally, shop drawings will showcase the materials selected and method statement which includes the steps proposed for construction. An example of items to be included in a flooring method statement is shown in Table 4.2. Shop drawings indicate dimensions, materials, finishes and details associated with their installation to ensure that fabrication will be accurate as what has been specified. Moreover, shop drawings have to be approved and certified by a design professional to ascertain that the products and systems are in compliance with the quality standards of the contract documents. This is an important procedure as an error in the fabrication, especially long lead time items, can put the workmanship quality in jeopardy as contractors will then have to neglect on the quality aspect in order to meet the construction deadline (Emmitt and Gorse 2010).

Moreover, the design of certain items has to be practical to ensure that owners are able to take care of the area with an acceptable level of effort. One example is if the flooring outside the toilet is made of timber, wider stepping piece should be designed to prevent water staining on the timber flooring. Another example is that shop drawings should also be checked to ensure that M&E protrusion are capped in order to hide untidy joints, a sign of poor workmanship quality. The specification of epoxy grouting in the floor plan drawing is also essential to ascertain that grouting will be used for tiles to be jointed smoothly, creating better aesthetics appearance and enhancing workmanship quality as well. Therefore, a rigorous

Table 4.2 Items in flooring method statement

No.	Items in flooring method statement
1	Flooring system
2	Method of installation
3	Type of adhesives
4	Type of finishing coat
5	Provision for movement (sufficient expansion gap and/or staggered joints)
6	Surface preparation
7	Preparation and laying of screed
8	Preparation and laying of sub-base
9	Laying flooring
10	Sanding
11	Applying finishing coats
12	Process of skirting installation

Source BCA (2012)

shop drawing submittal check is as critical as the pre-selection of construction materials before actual construction to pre-evaluate the resultant quality level and ascertain the success of a high CONQUAS score.

4.6.2 Constructing Mock-Ups to Check for Implications with Other Trades of Works: CM2

A clearer illustration is actually to construct site mock-ups to determine the implications of the proposed construction method and suitability of materials with other trades. Mock-ups require the contractor to build small models utilising the approved materials, either on site or in the factory. In a mock-up, specific nuances such as colours, shades, tones, patterns and textures of an installation can be better observed and inspected rather than just by referring to paper documents (Jackson 2010). In addition, the mock-ups will have to be evaluated to ensure its compliance and understanding of the workmanship quality with each manufacturer's requirements after assembling it with other materials. On top of that, these representative assemblies will have to be acceptable by standards indicated in the contract documents as well as approved by the client.

However, mock-ups are not practical for every installation and contractors tend to do this only if they are specified in the contract documents (Meier and Wyatt 2008). But, it should be noted that mock-ups are extremely useful when the expectation for quality cannot be easily conveyed in the specifications. As reinforced by Kudder and Erdly (1998), one important CSF to confirm the quality expectation is to construct a mock-up as it would demonstrate an installation as similar as possible to the actual installation to foresee any construction difficulties. Contractors would then be able to determine the best method of construction which would not impact on the workmanship quality of the constructed product.

4.6.3 Field Demonstration by Labourers to Showcase Their Understanding of the Workmanship Quality Required: CM3

During the actual building construction phase, the workmanship quality of the labourers should be further inspected by requesting them to conduct field demonstration on a small portion of work first before constructing the rest. This is to prepare them for coordination with other subcontractors and communication between different trades to show the importance of following the construction sequence, having wet or dirty trade completed before dry trades begin to ensure workmanship quality (Tattersall 1990). Conducting field demonstration will also show the interrelationships of materials and critical construction processes to establish the quality standard by which the works will be evaluated.

Besides the assurance of having a skilled workforce from the subcontractors, this is an additional quality procedure which will ensure that the approved method of construction is actually employed successfully. They will be under the watchful eyes of the main contractor to make sure that their workmanship truly shows that they understand the CONQUAS standard required which will also help to secure a higher CONQUAS score.

4.6.4 Conducting Preparatory Inspection Using Template Checklist at Every Stage of Work Activity: CM4

The purpose of preparatory inspection is to discover patent defects by the reasonable exercise of due diligence at each step of the installation to avoid any errors and reworks that can disrupt the construction workmanship quality (Jackson 2010). Preparatory inspection has to be designed to check progress and make sure that everything is ready at every stage of construction. This inspection process is very different from inspecting the entire installation only but rather, a small area of installation would be pre-inspected, noting any deficiencies or variations from the specified quality indicated in the specifications. Photographs can also be taken as proof of defects compliance or non-compliance. This will ensure a high standard of workmanship quality and measures to preclude such recurrence should be employed (Ramsey 1984). This will also ensure higher chances of achieving superior CONQUAS scores as adjustments and corrections would have to be made to the satisfaction of the contract requirements before proceeding with the full installation process. However, not all contractors conduct preparatory inspections (Jackson 2010), but the findings suggest that they should because it costs a lot less to correct an error or defect early in the work sequence than it does after the system is completed (Rommel 1996). Examples of such template checklists can be referenced from what the BCA assessors are checking on as shown in Tables 4.3 and 4.4.

Table 4.3 Template for aluminium window inspection

In-process assessment	Activities to be carried out during on-site verification
Preparation	Check the size of the structural opening against the window dimensions. Ensure the size is within allowable tolerance and the correct type of window
	Set the datum line from finished floor level. Ensure the setting out of datum line and opening is correct
Installation of outer frame	Place the frame into the opening and secure by temporary means
	Check the plumb of the frame. Ensure the plumb (X- and Y-axis) is within allowable tolerance
	Check the setting out of the frame. Ensure the position (including level and horizontal alignment) of the frame is within allowable tolerance
	Before ramset the fish-tail brackets into the concrete structure, ensure the spacing of all brackets are within allowable tolerance including the edge distance of the 1st bracket
	Ensure that all brackets are properly fixed to the wall
	Ensure the gap between frame and wall/column/beam is within allowable tolerance
	Grout the gap between the frame and opening with cement mortar. Ensure no gap or void in the grouting and then the grouting is tooled to give neat surface or edge
	Upon curing of grout, apply cementitious waterproofing slurry on the external joint areas. Apply the slurry by using a hard plastic bristled brush to ensure consistency. Prior to application, the surface should be damp and free from debris/dust
	After completing the installation, ensure the protection tape is in good condition
Installation of inner frame	Upon completion of external painting, remove protection tape for glazing installation
	Ensure the correct type and size of frame is used
	Ensure the correct friction stay (for casement/top hung window) or safety screw/device (for sliding window) is used. The screws for securing the friction stay must be stainless steel
	Apply approved sealant caulking at window perimeter on the external surface prior to glass installation
Glazing	Ensure the correct type, size and thickness of glass is used
	Remove window beading and insert glass. Secure the glass with approved rubber gasket or sealant. Ensure the gasket is tight fitting and no gap at corners of gasket
	Ensure full protection to the glazing

Source BCA (2012)

Table 4.4 Checklist for internal wet-area waterproofing system inspection

In-process assessment	Internal wet-area waterproofing system inspection template checklist
Surface preparation	Ponding test should be conducted on bare concrete slab for 24 h
	The substrate surface should be smoothly finished. Surface irregularities must be properly filled with appropriate materials before the application of the waterproofing system
	The substrate surface must be free from sharp protrusions which may tear or punch through the waterproofing membrane
	The substrate surface must be cast to fall in the right direction so that water will not stagnate and pond on the slab. The stagnation of water allowable shall not cover a 20-cent coin when placed in the ponded water
	For toilets with masonry walls, the mortar joints should be flush pointed and rendered with cement/sand mix to a height of at least 300 mm above the floor level to receive the membrane. For the shower area, the walls should be rendered to at least 1,800 mm height
	Angle fillet should be provided at the floor/wall and floor/protrusion joints. Alternatively, a strip of fibreglass mat, serves as reinforcement, can be applied during application of membrane. Please specify which method to be used
	The substrate surface is cleaned and free from all contaminants like mud, dust, loose particles, cement laitance, etc.
	The concrete surface needs to be kept damp but not wet with clean water before application of waterproofing membrane
Application	The components of the waterproofing membrane shall be mixed according to the manufacturer's specifications
	When mixing the two components, the powder and the liquid, always add the powder into the liquid slowly. Mix with a low speed mechanical stirrer until a uniform and lump-free slurry is formed
	At least two coats of cementitious slurry are required
	During application, the waterproofing membrane shall not be contaminated or disturbed by external agents like rain water, leaking water or debris
	When using a roller, each coat of slurry must be applied at right angles to the previous coat
	At pipe protrusions in slabs, the membrane is applied up to finished floor level
	At gully areas, the membrane is turned down 50 mm into the gully trap
	The waterproofing membrane should turn up every wall at a height of 300 mm
	The waterproofing membrane should extend on horizontal dry surfaces, e.g., doorway, by 150 mm from edge of wet area
	At bath and shower area, the waterproofing membrane should be applied to at least 1,800 mm height and 1,500 mm width of the wall or the entire width of the enclosure
	Manufacturer's stated pot life shall be strictly followed
	The thickness of the membrane shall be verified
	The lag time between the application of the next coat and the previous coat must follow strictly the manufacturer's instructions s o that one coat is cured before the next coat is applied

Source BCA (2012)

Overall, the concept behind preparatory inspection is to inspect the work as it progresses, instead of waiting until it is completed, only to find that the quality is unacceptable. This preparatory inspection accomplishes two goals: first, the subcontractor is now well aware of the quality standard expected, and therefore, the end result should meet the expected quality (Saarinen and Hobel 1990); second, the cost to correct the deficiencies can be kept to a minimum (Rommel 1996). Inspection is thus the most direct way to effectively manage workmanship quality on those doing the work. This can be done by making use of specially tailored quality inspection checklists for every trades of each project. Drawing up checking procedures to prepare for CONQUAS assessment will highlight possible site problems that may affect workmanship quality (Hernandez and Aspinwall 2008). These checklists will be used for inspection when materials are delivered to site as well as throughout the whole construction process to verify the quality standards.

Unfortunately, there are many projects whose pace is so fast and furious that the project team may only conduct final inspection, let alone these preparatory ones (Jackson 2010). The idea behind preparatory inspection as a CSF is to catch everything that does not comply with the workmanship standards before the next subcontractor takes over. This is not only the responsibility of the entire project team but also the client and those in the field to enforce and execute the quality inspection task to the attainment of high CONQUAS scores.

4.6.5 Using Recognised Testing Standards to Check for Non-compliance During Inspection: CM5

There are a number of quality standards associated with the testing and manufacture of various products which must be met by the materials and equipment used in the project as spelled out in the contract specifications (Meier and Wyatt 2008). In Singapore, the commonly used standards are the British Standards or Singapore Standards, and contractors have to follow these specified criteria in order to achieve quality standards. Products have to meet the standards of these certifications to ensure they are in compliance and able to achieve the minimum quality standards. Otherwise, they will have to be removed and replaced to the satisfaction of the client and in the hope of achieving high CONQUAS scores as well.

4.6.6 Cross-Checking by Another Party: CM6

The industry have voiced that some companies require the project manager to cross-inspect and evaluate another project's site conditions of their company on a monthly basis to further detect any non-compliance to increase the chances of achieving a higher CONQUAS score during the actual BCA CONQUAS assessment exercise. Cross-checking by another party such as another supervisor, subcontractor, client or the architect will be more effective than checking done by someone who has been working on the project as areas of non-compliance are more visible to the eyes of a third party, who will be able to identify defects impartially.

To further increase the possibility of achieving high CONQUAS scores, sub-contractors, foremen and labourers should also report any quality issues to the main contractor at any point in time instead of relying solely on the inspection process. The problems identified would then be notified to the appropriate parties to get the problems fixed, thus enhancing the workmanship quality further.

4.6.7 Adherence to Reporting and Follow-Up Procedure of Defects Before CONQUAS Assessment: CM7

In the event where non-conforming work is discovered, it should be duly reported, and follow-up procedures should be executed until it meets the stipulated quality requirements (Coffey et al. 2011). This is essential as part of construction progress monitoring and process reporting to check for compliance with the quality standards. Defects will then have to be corrected and prepared for retesting again to ensure that the required workmanship quality standard is reached before actual CONQUAS assessment. For example, during the internal water ponding test, if any leakage is detected, remedial actions should first be taken before the re-ponding test is conducted by the BCA assessors.

However, such quality reporting and procedures involve reworks, warranty costs and control charts to identify quality problems and provide information on areas of possible improvement (Lockamy 1998). Hence, to minimise such costs of poor quality, follow-up procedures should be immediate so that corrective actions can be taken before defective products are produced (Flynn et al. 1995; Ho 1999). Therefore, the target of reaching a high CONQUAS score can be achieved as defects' non-compliance is eliminated before the actual CONQUAS assessments are conducted by the BCA assessors.

4.7 Summary of Critical Success Factors

Above all, as the quality benchmark is constantly increasing every year, contractors must also strive to do better and work hard to improve their quality management practices and endeavour to achieve a higher CONQUAS score and not just the minimum requirements. Otherwise, they will still be unable to meet the minimum CONQUAS requirements with the continuous development in the CONQUAS system. With all these 33 CSFs in place, the route to the achievement of the CONQUAS standards will be sufficient or even more than enough to realise the goal of attaining a high CONQUAS score. Table 4.5 summarises the 33 CSFs identified in the study.

Table 4.5 Critical success factors for achieving high CONQUAS scores

CSFs	*Human Resource Management*
HRM1	Formation of QMU
HRM2	Commitment of top management
HRM3	Supportive and cooperative employees
HRM4	Training employees
HRM5	Supervision by client or architect
HRM6	Learning from past experiences
HRM7	Benchmarking
CSFs	*Subcontract Management*
SCM1	Control number of subcontractors
SCM2	Rigorous pre-qualification process
SCM3	Skill level of labourers
SCM4	Collaborative efforts between subcontractors and main contractor
SCM5	Adequacy of contract period and contract sum
SCM6	Subcontractor liable for achieving a specific CONQUAS score
SCM7	Award incentives
SCM8	Impose penalty
SCM9	Specify clear instructions
CSFs	*Schedule Management*
SM1	Prefix sample locations
SM2	Adequacy of samples
SM3	Arrange for phased construction
SM4	Plan to book CONQUAS assessment to tie in with site schedule progress
CSFs	*Material Management*
MM1	Comprehensive materials selection process
MM2	Select materials which can be better managed during construction
MM3	Inspect materials upon delivery
MM4	Proper materials handling and storage
MM5	Protection of constructed components
MM6	Sample testing through Independent Testing Agency (ITA)
CSFs	*Construction Management*
CM1	Review of shop drawings
CM2	Construct mock-ups
CM3	Field demonstration
CM4	Preparatory inspection
CM5	Reference to recognised standards
CM6	Cross-checking
CM7	Follow-up defects reported

To sum up, it is observed that the 33 CSFs are the planning inputs required before actual construction begins. These are the basic and necessary components in carrying out a construction process, and the output of the process will be excellent workmanship quality for the achievement of high CONQUAS scores. Even so, it should be noted that during the construction process, additional quality management steps are still required to enhance the probability of reducing any form of non-compliance. It is also noted that some of the factors may be further classified

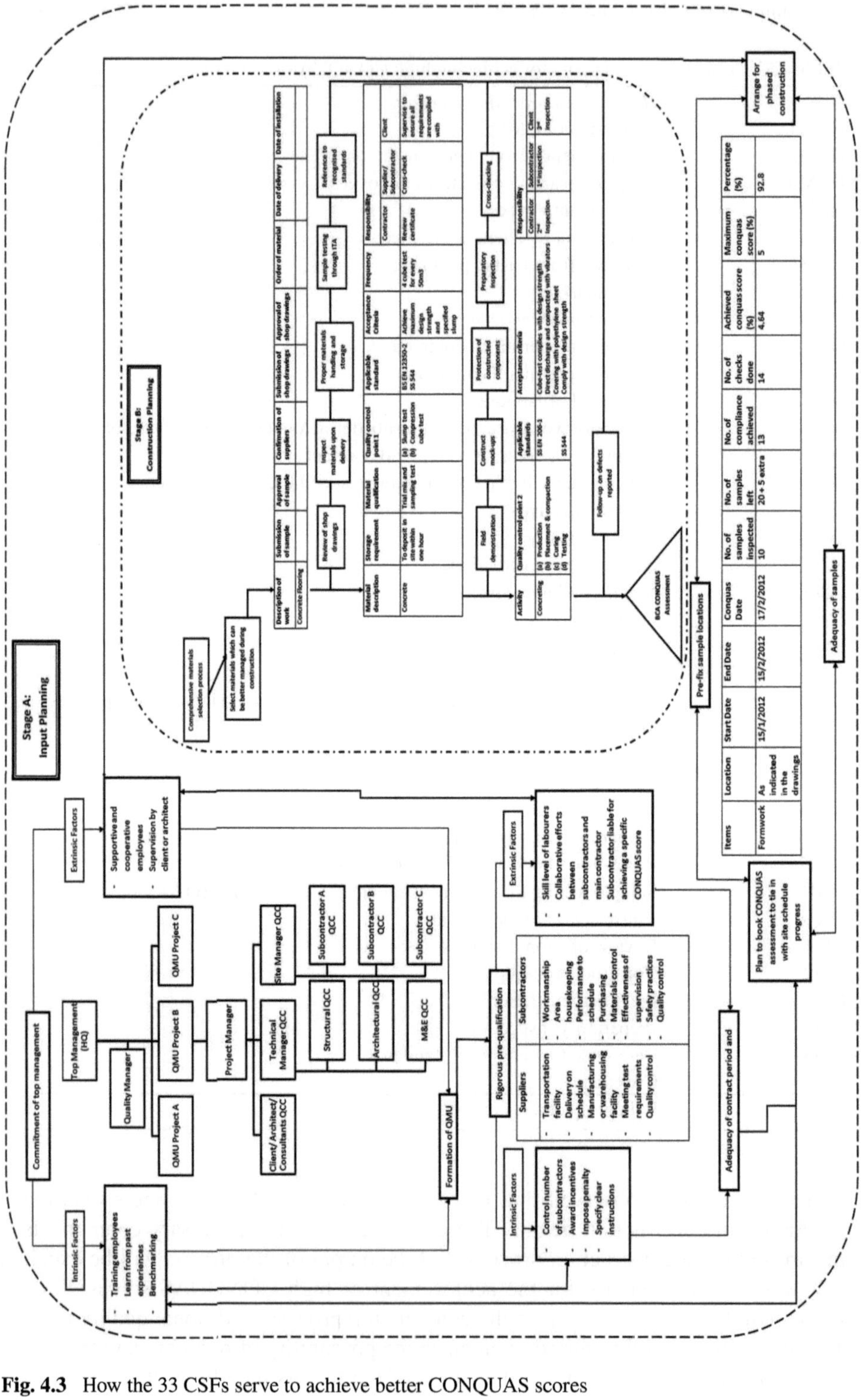

Fig. 4.3 How the 33 CSFs serve to achieve better CONQUAS scores

into intrinsic and extrinsic factors. Intrinsic factors are those CSFs which originate from the contractor's firm itself without any influence by extrinsic factors which emanate from other players of the project as well as due to factors which are not within the control of the main contractor. All that said, a conceptualised CONQUAS management framework is developed as shown in Fig. 4.3, which illustrates how the 33 CSFs serve to achieve high CONQUAS scores.

With that, this research is of the view that there are certain CSFs which are even more critical than others, and they have room for greater improvement which will help to enhance the CONQUAS score. Therefore, the next chapter will showcase the research methodology undertaken which attempts to further validate and differentiate among these CSFs—those that only help to achieve the necessary CONQUAS standards and those that are significant in realising the goal of a high CONQUAS score. Thereafter, this conceptualised framework will be further detailed in Chap. 8.

References

Abu-Taieh, E. M., El Sheikh, A. A., & Abu-Tayeh, J. M. (2009). Challenges in Implementing information technology plan. *Strategic Information Technology and Portfolio Management*. Jordan: Idea Group Inc.

Ansari, A., & Modarress, B. (1990). *Just-in-time purchasing*. New York: Free Press.

Anderson, J. C., Rungtusanatham, M., Schroeder, R. G., & Devaraj, S. (1995). A path analytic model of a theory of quality management underlying the deming management method: preliminary empirical findings. *Decision sciences*, 26(5), 637–658.

Ahire, S. L., & O'shaughnessy, K. C. (1998). The role of top management commitment in quality management: an empirical analysis of the auto parts industry. *International Journal of Quality Science*, 3(1), 5–37.

Ahire, S. L., & Dreyfus, P. (2000). The impact of design management and process management on quality: an empirical investigation. *Journal of Operations Management*, 18(5), 549–575.

Barrier, M. (1992). Small firms put quality first. *Nation's Business*, 80(5), 22–32.

Bell, R. R., & Burnham, J. M. (1989). The paradox of manufacturing productivity and innovation. *Business Horizons*, 32(5), 58–64.

Bennett, F. L. L. (2012). *The management of construction: A project lifecycle approach*. London: Taylor & Francis.

Brennan, D. S. (2008). *The construction contracts book: How to find common ground in negotiating the 2007 industry form contract documents*. American Bar Association.

Building and Construction Authority. (2005). *CONQUAS 21: The BCA construction quality assessment system*. Singapore.

Building and Construction Authority. (2010). *CONQUAS enhancement series*. Retrieved August 20, 2012, from http://www.bca.gov.sg/Publications/EnhancementSeries/enhancement_series.html.

Building and Construction Authority. (2012). *BCA construction quality assessment system CONQUAS application procedure*. Retrieved October 2, 2013, from http://www.bca.gov.sg/professionals/iquas/conquas_appli.html.

Burt, D. N. (1989). Managing product quality through strategic purchasing. *Sloan Management Review*, 30(3), 39–48.

Camp, R. C. (1989). Benchmarking: the search for industry best practices that lead to superior performance. In *Benchmarking: the search for industry best practices that lead to superior performance*. ASQC/Quality Resources.

Carmel, E. (1995). Cycle time in packaged software firms. *Journal of Product Innovation Management*, 12(2), 110–123.

Chapman, R. L., Clarke, P., & Sloan, T. (1991). TQM in continuous-process manufacturing: Dow-Corning (Australia) Pty Ltd. *International Journal of Quality & Reliability Management*, 8(5), 77–90.

Coffey, V., Willar, D., & Trigunarsyah, B. (2011). Quality management system and construction performance. In *Proceedings of: 2011 IEEE International Conference on Quality and Reliability*, IEEE Computer Society (pp. 14-17). Bangkok, Thailand.

Construction Industry Development Board. (1989). *Managing construction quality: A CIDB manual on quality management systems for construction operations*. Singapore.

Cole, C. C., Clark, M. L., & Nemec, C. (1993). Reengineering information systems at Cincinnati Milacron. *Strategy & Leadership*, 21(3), 22–48.

Corenet. (2012). *List of Projects Scored under CONQUAS*. Retrieved August 14, 2012, from http://www.corenet.gov.sg/homeowners/listsql/default.asp?SortBy=Q29udHJhY3Rvcg==&category=QUxMIENBVEVHT1JJRVM= PRIVATE%20HOUSING%20.

Daft, R. L. (1998). *Essentials of organization theory and design*. Cincinnati, Ohio: South-Western College Publishing.

Dainty, A. R., Millett, S. J., & Briscoe, G. H. (2001). New perspectives on construction supply chain integration. *Supply Chain Management: An International Journal*, 6(4), 163–173.

Dale, B. G., & Duncalf, A. J. (1984). A study of quality assurance in small businesses. In *Proceedings of the Institution of Mechanical Engineers, Part B: Journal of Engineering Manufacture*, (Vol. 198, pp. 135–139).

Dainty, A., & Loosemore, M. (2012). *Human resource management in construction*. London: Taylor & Francis.

Delgado-Hernandez, D. J., & Aspinwall, E. (2008). A framework for building quality into construction projects—Part I. *Total Quality Management and Business Excellence,19*(10), 1013–1028.

Deming, W.E. (1986). *Out of the crisis*. MIT Center for Advanced Engineering Study, Cambridge, MA

Din, S., Abd-Hamid, Z., & Bryde, D. J. (2011). ISO 9000 certification and construction project performance: The Malaysian experience. *International Journal of Project Management,29*(8), 1044–1056. doi:10.1016/j.ijproman.2010.11.001.

Dutton, J. E., & Ashford, S. J. (1993). Selling issues to top management. *Academy of Management Review*, 18(3), 397–428.

Ellram, L. M. (1991). Supply-chain management: the industrial organisation perspective. *International Journal of Physical Distribution & Logistics Management*, 21(1), 13–22.

Emmitt, S., & Gorse, C. (2010). *Barry's advanced construction of buildings*. New York: Wiley.

Eom, C., Yun, S., & Paek, J. (2008). Subcontractor evaluation and management framework for strategic partnering. *Journal of Construction Engineering and Management,134*(11), 842–851.

Everett, R. J., & Sohal, A. S. (1991). Individual involvement and intervention in quality improvement programmes: using the Andon system. *International Journal of Quality & Reliability Management*, 8(2), 21–34.

Fewings, P. (2005). construction project management: An integrated approach. London: Taylor & Francis.

Ferdows, K., & De Meyer, A. (1990). Lasting improvements in manufacturing performance: in search of a new theory. *Journal of Operations Management*, 9(2), 168–184.

Flynn, B. B., Sakakibara, S., & Schroeder, R. G. (1995). Relationship between JIT and TQM: practices and performance. *Academy of Management Journal*,38(5), 1325–1360.

Fong, S. W., Cheng, E. W. L., & Ho, D. C. K. (1998). *Benchmarking: a general reading for management practitioners*. Management Decision. Retrieved January 2, 2013, from http://202.120.148.188/rlzygl/llqy/Benchmarking%20a%20general%20reading%20for%20management.pdf.

Fouayzi, H., Caswell, J. A., & Hooker, N. H. (2006). Motivations of fresh-cut produce firms to implement quality management systems. *Applied Economic Perspectives and Policy*, 28(1), 132–146.

Griffith, A. (2011). *Integrated management systems for construction: Quality, environment and safety*. Trans-Atlantic Publications.

Grieco, P. L., & Gozzo, M. W. (1995). *Total Quality Management: Just in Time*. Kuala Lumpur.

Ghobadian, A., & Gallear, D. N. (1996). Total quality management in SMEs. *Omega*, 24(1), 83–106.

Harber, D., Marriott, F., & Idrus, N. (1991). Employee participation in TQC: an integrative review. *International Journal of Quality & Reliability Management*, 8(5), 24–34.

Hernandez, D. J., & Aspinwall, E. (2008). Quality management case studies in the UK construction industry. *Total Quality Management*, 19(9), 919–938.

Ho, S. (1999). *Operations and quality management*. Cengage Learning EMEA.

Ho, D. C., Cheng, E. W., & Fong, P. S. (2000). Integration of value analysis and total quality management: the way ahead in the next millennium. *Total Quality Management*, 11(2), 179–186.

Jackson, B. J. (2010). *Construction management JumpStart: The best first step toward a career in construction management*. New York: Wiley.

Jelinek, M., & Adler, N. J. (1988). Women: World-class managers for global competition. *The Academy of Management Executive*, 2(1), 11–19.

Juran, J. (1988). *Juran on planning for quality*. American Society for Quality Control, Milwaukee, WI.

Kanji, G. K., & Wong, A. (1998). Quality culture in the construction industry. *Total Quality Management,9*(4–5), 133–140.

Keeble, T. (2006). *The modern period room 1870–1950: The construction of the exhibited interior*. London: Taylor & Francis.

Koehn, E., & Regmi, D. (1990). Quality in constructed projects: International firms and developing countries. *Journal of Professional Issues in Engineering,116*(4), 388–396. doi:10.1061/(ASCE)1052-3928(1990)116:4(388).

Kouzmin, A., Löffler, E., Klages, H., & Korac-Kakabadse, N. (1999). Benchmarking and performance measurement in public sectors: towards learning for agency effectiveness. *International Journal of Public Sector Management*, 12(2), 121–144.

Kudder, R. J., & Erdly, J. L. (1998). *Water leakage through building facades*. ASTM.

Kumaraswamy, M. M., & Matthews, J. D. (2000). Improved subcontractor selection employing partnering principles. *Journal of Management in Engineering*, 16(3), 47–57.

Langford, D. A., & Rowland, V. R. (1995). *Managing overseas construction contracting*. Thomas Telford.

Leach, N. R., Krajewski, L. J., & Ritzman, L. P. (1993). *Operations Management: Study Guide*. Prentice Hall.

Levy, S. M. (2009). *Construction process planning and management: An owner's guide to successful projects*. Butterworth-Heinemann.

Linklow, P. (1989) Is your culture ready for total quality?. *Quality Progress*, 22(11), 69–71.

Lockamy III, A. (1998). Quality-focused performance measurement systems: a normative model. *International Journal of Operations & Production Management*, 18(8), 740–766.

Matthews, J. D., Tyler, A., & Thorpe, T. (1996). Subcontracting: the subcontractor's view. *The organization and management of construction: shaping theory and practice* (Vol. 2, pp. 471–480), London, UK: E & FN Spon.

Meier, H. W., & Wyatt, D. J. (2008). *Construction specifications: Principles and applications*. Thomson Delmar Learning.

Mincks, W. R., & Johnston, H. (2004). *Construction jobsite management*. Thomson Delmar Learning.

Ogunlana, S. (2003). *Profitable Partnering in Construction Procurement*: Taylor & Francis.

Palaneeswaran, E., Ng, T., & Kumaraswamy, M. (2006). Client satisfaction and quality management systems in contractor organizations. *Building and Environment,41*(11), 1557–1570. doi:10.1016/j.buildenv.2005.06.004.

Peurifoy, R., Schexnayder, C. J., Shapira, A., & Schmitt, R. (2010). *Construction planning, equipment, and methods*. New York: McGraw-Hill Companies, Incorporated.

Pike, R. J., Barnes, R., & Barnes, R. J. (1995). *TQM in Action: A practical approach to continuous performance improvement*. Springer.

Ramsey, T. (1984). Quality control "A Necessity Not an Option". *Journal of Construction Engineering and Management,110*(4), 513–517. doi:10.1061/(ASCE)0733-9364(1984)110:4(513).

Robert, A. O., & Linda, L. B. (2000). An integrated view of project and quality management for project-based organizations. *International Journal of Quality and Reliability Management,17*(4), 351–363.

Rommel, G. (1996). *Quality pays*. New York: Macmillan Press.

Rounds, J. L., & Segner, R. O. (2010). *Construction supervision*. New York: Wiley.

Saarinen, A., & Hobel, M. (1990). Setting and meeting requirements for quality. *Journal of Management in Engineering,6*(2), 177–185. doi:10.1061/(ASCE)9742-597X(1990)6:2(177).

Saraph, J. V., Benson, P. G., & Schroeder, R. G. (1989). An instrument for measuring the critical factors of quality management. *Decision Sciences*, 20(4), 810–829.

Shin, H., Collier, D. A., & Wilson, D. D. (2000). Supply management orientation and supplier/buyer performance. *Journal of Operations Management*, 18(3), 317–333.

Sommerville, J., & Sulaiman, N. F. (1997). The culture for quality within the UK construction industry: temporal relatedness and dominance. *Total Quality Management*, 8(2-3), 279–285.

Tan, K. C. (2001). A framework of supply chain management literature. European Journal of Purchasing & Supply Management, 7(1), 39–48.

Tabassi, A. A., & Bakar, A. H. A. (2009). Training, motivation, and performance: The case of human resource management in construction projects in Mashhad, Iran. *International Journal of Project Management,27*(5), 471–480.

Tattersall, G. H. (1990). *Workability and quality control of concrete*. London: Taylor & Francis.

Thorpe, B., & Sumner, P. (2004). *Quality management in construction*. Gower: Gower Publishing Ltd.

Trent, R. J., & Monczka, R. M. (1999). Achieving world-class supplier quality. *Total Quality Management*, 10(6), 927–938.

Thorpe, B., Sumner, P., & Duncan, J. M. (1996). *Quality assurance in construction*, Gower.

Westbrook, J. D. (1993). Organizational culture and its relationship to TQM. *Industrial Management-Chicago Then Atlanta*, 35, 1–1.

Wilson, L. A., & Durant, R. F. (1994). Evaluating TQM: the case for a theory driven approach. *Public Administration Review*, 54(2), 137–46.

Wilson, D. D., & Collier, D. A. (2000). An empirical investigation of the Malcolm Baldrige National Quality Award causal model. Decision Sciences, 31(2), 361–383.

Wood, G. D., & Ellis, R. C. (2005). Main contractor experiences of partnering relationships on UK construction projects. *Construction Management and Economics*, 23(3), 317–325.

Zeitz, G., Johannesson, R., & Ritchie, J. E. (1997). An employee survey measuring total quality management practices and culture development and validation. *Group & Organization Management*, 22(4), 414–444.

Chapter 5
Research Design and Methodology

5.1 Overview of Research Process

Based on empirical findings conducted by others in the extensive literature review as well as logical deductions, because achieving "high" CONQUAS score is not an obligatory rule and since it is hard to realise too, few contractors do their best to meet this higher target. Most of them appear to simply do what is needed to meet the minimum quality requirements as dictated by regulations (Low and Omar 1997). Hence, the research process will now turn to do a study on all the 33 CSFs using a three-tier field investigation strategy.

In the first stage, survey would be conducted mainly to sieve out which CSFs are the "Basic Attributes" and which are the "Features". Basic attributes are the fundamental factors (OECD 2002) which are necessary to achieve the minimum CONQUAS standards, whereas features are the additional factors that are good to have (Riebisch 2004) so as to aid in enhancing the CONQUAS scores. Contractors who already have the "Basic Attributes" in place can then know which "Features" to prioritise on and apply in their CONQUAS management workflow so as to perfect their workmanship quality. The effectiveness of this approach which was used by many researchers such as (Walker and Chau 1999), Wang et al. (2003) and many others has convinced this research to adopt a similar tactic. This is because many of the 33 CSFs cover qualitative items such as management efforts for which hard performance data are not available.

Following that, interviews would be carried out with industry professionals to ascertain the findings of the survey. Both the survey and interview also serve to validate the seven sets of research hypotheses set out in Chap. 1.

Lastly, a real-life case study would be examined to further support the responses received from the survey and interviews by investigating a contractor's firm whose CONQUAS scores have improved tremendously over the years to find out the change in attitude and practices, as well as the reasons and motivation for doing so which contributed to such a great improvement in their workmanship quality standard.

S. P. Low and J. Ong, *Project Quality Management*, DOI: 10.1007/978-981-287-074-2_5,

Overall, with this three-tier research methodology, this research would then be able to justify the proposed CONQUAS management framework in helping contractors to strategise their focus area to achieving high CONQUAS scores.

5.2 Survey Design

The purpose of this survey is to distinguish the importance level and extent of adoption of the 33 CSFs in influencing and affecting the CONQUAS scores. It is deemed that the best approach would be to garner the perceptions of contractor's firms who are required to abide by the CONQUAS standards as well as pursue quality excellence to gain recognition from the industry. Survey was chosen because it is good for generating quantitative data Zillmer et al. (1990) and enabling a statistical analysis to pick out the two groups of CSFs.

First, a pilot survey was conducted mainly to gather feedbacks and comments to the structure of the questions with respondents recognising the objective of each question. This is to enhance the accuracy of the survey, having questions that are correctly designed to serve their intended purposes. This serves to ensure a better understanding of the factors in the survey and decrease the chance of misinterpretation. Following the pilot study, the survey questions were revised and ready for a full-scale survey to be conducted. Blank space was available for respondents to provide their comments and/or suggest their own factors which were not covered in the question.

5.2.1 Type of Respondents

First and foremost, respondents have to be well versed in the CONQUAS standards and have the experience in managing CONQUAS building projects. Hence, the project manager, quality manager, engineer or any other relevant personnel who are equipped with CONQUAS knowledge would be the targeted respondents for this survey. This ensures that the sample group will consist of a good mix of different types of professionals and practitioners in the building industry, making it a multidisciplinary combination for this survey to yield a representative result. They would be able to provide insights into the 33 CSFs identified and make a stand in determining the importance of each factor in meeting a high CONQUAS score as well as whether the factors were actually being implemented in practice.

Correspondences were sent via email to all 62 A1, 24 A2, 28 B1 and 5 B2 general building contractors who have undertaken at least one CONQUAS project to seek their participation in this survey. The classification of the various financial grades is explained in Table 5.1. This sample population was chosen as only these four categories of contractors must be ISO 9001:2008 accredited, which means that they have an adequate knowledge on having a QMS in place which

Table 5.1 Financial grade classification of the general building contractor

Grade	Financial (in million) (min paid-up capital and min net worth)	Management and development	Track record (in million) (past 3 years)	Additional requirements
A1	S$15	(1) 24RP/P/T and minimum 8RP (2) ISO 9001:2008 (SAC) (3) ISO 14000 (4) OHSAS 18000/SS 506 Part 1	S$150.0 of which - S$75.0 PS - S$112.5 MC - S$37.5 SP	(1) Annual submission of financial accounts and certified VAP calculation (2) Annual submission of CET declaration—for A1, A2 (3) General Builder Licence – Class 1 (GB1) (4) At least one RP/P/T with CCPM—for A1, A2 (5) At least one RP/P/T with CCPM—for B1 (6) At least one RP/P/T with BCCPE—for B2
A2	S$6.5	(1) 12RP/P/T and minimum 4RP (2) ISO 9001:2008 (SAC) (3) ISO 14000 (4) OHSAS 18000/SS 506 Part 1	S$65.0 of which - S$32.5 PS - S$48.75 MC - S$16.25 SP	
B1	S$3	(1) 6RP/P/T and minimum 2 RP (2) ISO 9001:2008 (SAC) (3) OHSAS 18000/SS 506 Part 1 (4) ISO 14000	S$30.0 of which - S$22.5 MC - S$7.5 SP - S$10.0 of which	
B2	S$1	(1) 3RP/P/T and minimum 1RP (2) ISO 9001:2008 (SAC) (3) OHSAS 18000/SS 506 Part 1 (4) ISO 14000	S$10.0 of which - S$7.5 MC - S$2.5 SP	

(continued)

Table 5.1 (continued)

Grade	Financial (in million) (min paid-up capital and min net worth)	Management and development	Track record (in million) (past 3 years)	Additional requirements
C1	S$0.3	(1) 1P + 1T (2) SMC	S$3.0	(1) GB1 or GB2 (2) At least one RP/P/T with BCCPE
C2	S$0.1	(1) 1P or 2T (2) SMC	S$1.0	
C3	S$0.025	(1) 1T	S$0.1	

Note PS projects executed in Singapore, *MC* main contracts (nominated sub-contracts may be included), *SP* minimum size single project, *RP* professional with qualifications recognised by Professional Engineers Board (*PEB*) of Singapore, Board of Architects (*BOA*) of Singapore or BCA (recognised for resident engineer), *P/T* professional and technical personnel with relevant qualifications, *SAC* Singapore Accreditation Council, *SMC* Safety Management Certificate, *CCPM* Certificate Course in Construction Productivity Management conducted by BCA Academy, *BCCPE* basic concept in construction productivity enhancement (certificate of attendance) conducted by BCA Academy

Source (BCA 2012)

will wincrease the accuracy of the information gathered. Also, research findings have shown that ISO 9001 certification provides a stepping stone towards quality management practices (Quazi and Padibjo 1998).

Although it is considered beneficial to have a greater sample size and at least 30 respondents, it is more important that the company elects a suitable respondent who is experienced in the scope of the survey. Additionally, in order to minimise the possibility of biased responses in the survey due to the different financial scale of firms the respondents belong to, this research aims to get at least 30 % (average response rate) of responses from each of these four groups (A1, A2, B1 and B2 main contractors). This is to ensure a well-balanced mix of respondents. As a result, the entire duration of the survey took about three months, from September 2012 to November 2012 to obtain the required sample size.

5.2.2 Survey Questions

To obtain a clearer picture of the intention of the survey, the survey questionnaire which is divided into three sections will be discussed in further details. A sample of the survey questionnaire is found in Appendix A.

In the first section, the objective is to find out the demographic profile of the respondent and background of his or her company with regard to managing CONQUAS projects. This information would be useful to explain the data collected.

As the literature review has already established the need for the existence of all 33 CSFs, they are automatically deemed to at least qualify as "Basic Attributes". Hence, the second section of this survey questionnaire used a Likert scale of 1–5 to rate the importance level of all the 33 CSFs in influencing the achievement of high CONQUAS scores. A mean rating for each of the CSFs would then be tabulated and refined such that those with a mean score of 4.00 and above are to be shifted to the "Features" group, considering that the value 4 in the 5-point Likert scale implies that the factor is important, a reference point used by Lu et al. (2008) as well.

Unlike the above section which is relatively subjective, the third section of the survey questionnaire is more straightforward and objective to reduce any chances of biased data. This question aids in assessing how the adoption of a particular CSF affects the CONQUAS score quantitatively. Basically, this is a "Yes" or "No" question, whereby respondents were asked to indicate those CSFs which they implement in most of their CONQUAS projects and those who answered "Yes" will be given a value 1 and "No" answers will have a value of 0. The factors which have a relatively low usage mean rating, below 0.5, and are in the "Features" group, imply that contractors would probably have to work harder on these areas so as to improve their workmanship quality and achieve higher CONQUAS scores.

Overall, the three-part structure of the survey questionnaire would enable an in-depth analysis of the 33 CSFs identified in Chap. 4.

5.2.3 Methodology for Survey Analysis

Data collected from the survey were analysed using IBM Statistical Package for Social Science (SPSS) version 21 as well as using charts and graphs from the Microsoft Office to illustrate and compare data. Descriptive statistics, including the mean ratings, percentages and standard deviations, were also computed. This forms the basis for the subsequent inferential quantitative data analysis and allows for greater understanding of the data trends.

Next, a series of parametric t tests was employed to justify the assumption that samples collected are representative of the population. This would ensure accuracy of the respondents' perceptions of the CSFs and allows generalisation to be made. The one-sample t test was used to examine the statistical significance of the mean importance ratings of the CSFs as well as the mean usage level of the CSFs. The coefficient of determination (R^2) was also examined to check whether there is any possibility that the degree of usage of the CSFs is influenced by its importance level and vice versa.

The Pearson's correlation analysis was then undertaken to study the correlations between each CSF variable and every other CSF variables to aid in validating the proposed CONQUAS management framework by establishing the links and relationships between the CSFs. The Pearson's correlation coefficient was chosen as it is appropriate for measurements that have an interval scale which are meaningful. Correlation summarises the strength of relationship between two variables, but it is important to remember that correlation is not causation. For this study, a positive correlation is one in which the ratings of both variables increase together and a negative correlation is one in which the ratings of one variable increase as the other variable decrease. A perfect correlation of exactly $+1$ or -1 will arise if the relationship between the two variables is exactly linear. If the correlation is 0, there is no linear relationship between the two variables and they are said to have no correlation; they are completely random.

Based on the literature review, a key research problem identified is the lack of quantification in determining the exact increase or decrease in CONQUAS scores with the application of each and every CSF. As the implementation of all CSFs may require more time, cost and manpower, it may not be worthwhile to do so just to achieve high CONQUAS scores. Hence, it is essential that an adequate methodology be set up via a linear regression technique to determine the extent of impact of the CSFs on the CONQUAS scores, how does the addition of a certain predictor (CSFs) improve the outcome (average CONQUAS score obtained in the recent three years) as measured by the increased in R^2. This is important to know so that contractors would be able to know the estimated CONQUAS score when they select only some CSFs to implement.

Therefore, multiple linear regressions were subsequently conducted as an analytical method to derive the relationship, strength and variability between the predictor (CSFs) and outcome variables (average CONQUAS score obtained in the recent 3 years). To use this regression technique, it is assumed that variables are normally distributed. Model equations would be presented and utilised as a reference for the proposed CONQUAS management framework.

Lastly, as the literature findings revealed that the financial capability of the contractors may have an influence over the degree of importance and usage level of the CSFs, independent t tests were also conducted to determine the validity of this proposition.

5.3 Interview Design

To enhance the validity of the quantitative results, qualitative research through interviews with three experts in the Singapore's building industry was carried out to investigate the results of the survey and also uncover relevant indicators, if any, that have been left out from the CSFs identified previously from the literature review. The objective of the three face-to-face interviews is to probe specific but dynamic questions that the quantitative survey is unable to address, to allow an understanding of how professionals perceived the 33 CSFs identified via open-ended questions rather than the 5-point Likert scale method and "Yes" or "No" questions employed in the survey questionnaire.

5.3.1 Interviewee's Profile

The experts were selected based on the type of organisation they belong to and their role in the company in managing CONQUAS to gather various opinions on the survey results as well as their approach to managing CONQUAS as shown in Table 5.2. The profile of the interviewees is also depicted to make known their background in the building industry, in particular their dealings with CONQUAS management. They were encouraged to give honest responses and not what the interviewer wants to hear to ensure a reliable qualitative data on the examination of the 33 CSFs. They were interviewed between November 2012 and December 2012.

5.3.2 Interview Questions

Based on the findings of the survey, the interview questions were drafted according to the survey responses of the contractors to garner more inferences and explanations. Particular attention was paid to responses that were contradictory to the literature review to provide insights into the inconsistent responses. Some questions may vary between the experts, depending on their company and individual profile as well as their role in managing CONQUAS to enable an in-depth understanding of their views on the 33 CSFs identified. The interview transcript for each of the three experts can be found in Appendices B, C and D, respectively. Findings from this interview would be used to explain the results of the survey in Chap. 6.

Table 5.2 Details of interviewee

Interviewee	Type	Background
Mr. A	Contractor's firm with good CONQUAS score track records	Mr. A graduated with a BSc (construction management) degree and is currently based in the firm's headquarter as a quality assurance manager. He is in charge of quality and productivity management for CONQUAS and ISO certification, buildability and constructability issues and to score well in BDAS. He is also responsible for implementing quality policies and overseeing organisational excellence at all site offices
Ms. B	Contractor's firm with poor CONQUAS score track records	Ms. B graduated with a BE (civil engineering) degree and is now a structural engineer. She is currently working on a redevelopment commercial project which is scheduled for CONQUAS assessment. Once this project is completed, she has been earmarked for another redevelopment commercial project slated for CONQUAS assessment as well. Besides handling structural engineering issues, her role also entails schedule management and supervising workmanship quality on site for all trades as well as arranging for CONQUAS inspection
Mr. C	Quality practitioner	Mr. C has been a quality practitioner for about 15 years and is well equipped with the quality and certification knowledge. He has also been involved in the development of CONQUAS over the years. In addition, he is also familiar with the approach that some firms undertake to managing CONQUAS

5.4 Case Study Design

The third stage of the research process is to understand a real-life phenomenon, using case study as a methodology to do a pre–post case study inquiry to examine a company (referred to as Company Z) at two points in time, that is, the period before and after the tremendous improvement in CONQUAS scores. This serves to further justify the usefulness of the 33 CSFs in helping to enhance the CONQUAS scores. Furthermore, the regression model identified in the survey analysis was used as a template to compare with the empirical results of the case study. If it is shown to support the same proposition, replication of this regression model may be claimed through this analytical generalisation. In this case study, besides getting an oral account from an experienced personnel who has undergone this transition phase, hard-copy evidence of documents was also referenced. This shows the research's attempt at trying to ensure that this case study is of sufficient rigour with multiple sources of evidence to back up claims for explanation to ensure that subjective judgements were not used to collect the data. A complete analysis of this case study is described in Chap. 7.

References

Building and Construction Authority. (2012). *Specific registration requirements for construction workhead*. Retrieved November 18, 2012, from http://www.bca.gov.sg/ContractorsRegistry/others/Registration_CW.pdf

Low, S. P., & Omar, H. F. (1997). The effective maintenance of quality management systems in the construction industry. *International Journal of Quality and Reliability Management, 14*(8), 768–790.

Lu, W., Shen, L., & Yam, M. C. (2008). Critical success factors for competitiveness of contractors: China study. *Journal of Construction Engineering and Management, 134*(12), 972–982.

Organisation for Economic Co-operation and Development. (2002). The value of fundamental research. Retrieved October 12, 2012, from http://iupab.org/publications/value-of-fundamental-research/

Quazi, H. A., & Padibjo, S. R. (1998). A journey toward total quality management through ISO 9000 certification-a study on small-and medium-sized enterprises in Singapore. *International Journal of Quality and Reliability Management, 15*(5), 489–508.

Riebisch, M. (2004). Supporting evolutionary development by feature models and traceability links. In Proceedings of the 11th IEEE International Conference and Workshop (pp. 370–377). Czech Republic.

Walker, A., & Chau, K. W. (1999). The relationship between construction project management theory and transaction cost economics. *Engineering Construction and Architectural Management, 6*(2), 166–176.

Wang, H., Zheng, Y. B., Liu, Z. Y., & Shen, X. K. (2003). DEI agent: an agent-based dynamic entity generating tool. *Machine Learning and Cybernetics International Conference, 1*, 535–539.

Zillmer, E. A., Fowler, P. C., Gutnick, H. N., & Becker, E. (1990). Comparison of two cognitive bedside screening instruments in nursing home residents: A factor analytic study. *Journal of Gerontology, 45*(2), 69–74.

Chapter 6
Results and Analysis

6.1 Overview of Research Analysis

A summary of the overall analysis process is provided to give an overview of the steps taken in this research to analyse the findings obtained as shown in Table 6.1.

6.2 Survey Response Rate and Demographics

Out of 119 survey questionnaires sent out, a total of 38 valid responses were received, giving a survey completion response rate of 32 %. The designation profile of the respondents is shown in Table 6.2. The mean number of years the respondents have spent in the industry is 11; the most experienced respondent spent 35 years in the industry even before CONQUAS was implemented, while the least number of years spent is 2.

To substantiate the validity of the responses received, it has been cross-verified that the sample responses received have undertaken a sizeable number of CONQUAS projects as compared to the sample population on average as shown in Fig. 6.1. It is also noted that the sample respondents' average CONQUAS score is slightly higher than the sample population except for B2 contractors as shown in Fig. 6.2.

6.3 Importance Level of the Critical Success Factors

The second section of the survey employs a 5-point Likert scale which aims to highlight the industry's perception of the importance level of each of the 33 CSFs and answer the research hypothesis 1 set out in Chap. 1 that there have to be certain "Features" which are more important than "Basic Attributes" in the attainment

S. P. Low and J. Ong, *Project Quality Management*, DOI: 10.1007/978-981-287-074-2_6,

Table 6.1 Overall structure of research analysis process

Type	Purpose of analysis process
Descriptive statistics (survey)	*Mean rating* To verify the importance level as well as the extent of adoption of the CSFs based on the results obtained
Inferential statistics (survey)	*Hypothesis 1* To show that there exist certain elements (features) that are more critical than the rest (basic attributes), having a greater influence on the CONQUAS scores which should be taken note of
	Hypothesis 2 To show that the different extent of adoption of the CSFs in the CONQUAS management workflow leads to the varying range of CONQUAS scores obtained
	Hypothesis 3 To show that there is no correlation between the importance level and extent of adoption of the CSFs through the coefficient of determination test
	Hypothesis 4 To show that there is correlation among the importance level of the CSFs
	Hypothesis 5 To show that there is correlation among the extent of adoption of the CSFs
	Multiple Linear Regression Model To determine the predicted CONQUAS score based on the CSFs implemented
	Hypothesis 6 To show that there is no distinction in the responses received from contractors of different building financial grades with regard to the importance level of the CSFs
	Hypothesis 7 To show that there is no distinction in the responses received from contractors of different building financial grades with regard to the extent of adoption of the CSFs
Qualitative evidence (interviews)	In the process of explaining the results of the descriptive and inferential statistics, findings from the three interviews conducted would be used to support the survey data analysis
Descriptive and exploratory (case study)	Using a real-life example, an account of a contractor's firm transition process, from the period when CONQUAS scores were low to the period when CONQUAS scores were comparatively higher, would be provided as well as used to validate the survey results and interview findings

Table 6.2 Designation profile of respondents

Designation	Frequency count	Percentage (%)
Project Manager	11	29
Technical Manager	12	32
Quality Assurance Manager	10	26
Others	5	13
Total	38	100

of high CONQUAS scores. The results obtained are shown in Table 6.3 and ranked in descending order based on the importance mean ratings.

From Table 6.3, one can infer from the mean scores that most respondents perceive "commitment of top management to advance the CONQUAS management performance (HRM2)", "getting the support and cooperation of employees (HRM3)" and "identifying a specific CONQUAS score that the subcontractor is liable for achieving (SCM6)" as the top three most important factors in influencing the

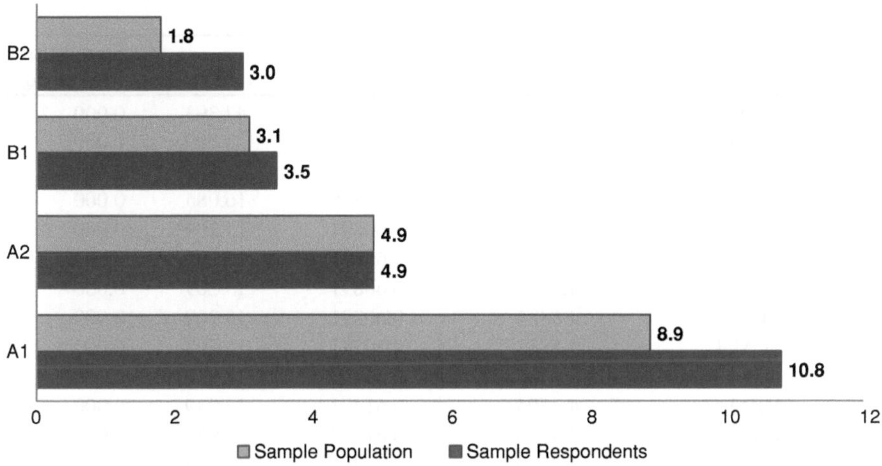

Fig. 6.1 Average number of CONQUAS projects undertaken by the different financial categories of contractors. *Source* BCA (2012)

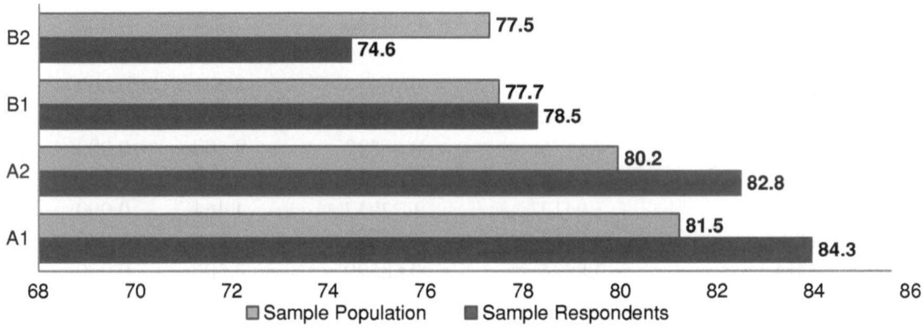

Fig. 6.2 Average CONQUAS scores obtained by the different financial categories of contractors. *Source* BCA (2012)

CONQUAS score. These three factors seem to suggest that dedication from all the various stakeholders in the project is essential, namely the main contractor's top management and employees as well as the subcontractors and labourers. This is because in the quality management effort, all members of an organisation should participate in improving the processes, services and culture in which they work (Wang 1998).

The importance of top management commitment is also seen in ISO 9001 where it is a requirement under Clause 5. Besides that, the top management needs the support of their employees to be involved in implementation and maintenance of the CONQUAS management framework (Leonard and Sasser 1982). This is because if commitment is not evident to and from the employees, then the CONQUAS management framework can never be truly implemented throughout the organisation. Employees at all levels of an organisation will not be committed to a framework that is not driven, supported and believed in by upper management

Table 6.3 Importance level of the 33 CSFs

Rank	Critical success factors	Importance mean rating	Standard deviation	*t* value	Significance (two-tailed)
1	HRM 2	4.9412	0.23883	47.393	0.000
2	HRM 3	4.7647	0.43056	23.899	0.000
3	SCM 6	4.7059	0.46250	21.507	0.000
4	SCM 9	4.6471	0.59708	16.085	0.000
5	HRM 1	4.5882	0.60891	15.209	0.000
6	HRM 7	4.5882	0.60891	15.209	0.000
7	SCM 3	4.5882	0.60891	15.209	0.000
8	MM 5	4.5882	0.60891	15.209	0.000
9	SCM 4	4.5882	0.70141	13.203	0.000
10	HRM 4	4.4706	0.61473	13.949	0.000
11	HRM 6	4.4706	0.61473	13.949	0.000
12	CM 7	4.4706	0.61473	13.949	0.000
13	CM 1	4.3529	0.77391	10.194	0.000
14	SM 1	4.2353	0.88963	8.097	0.000
15	MM 4	4.2353	0.74096	9.721	0.000
16	CM 4	4.2353	0.74096	9.721	0.000
17	SM 2	4.1176	0.84440	7.718	0.000
18	SM 4	4.1176	0.84440	7.718	0.000
19	SCM 7	4.1176	0.76929	8.471	0.000
20	MM 3	4.1176	0.76929	8.471	0.000
21	CM 2	4.1176	0.76929	8.471	0.000
22	MM 2	4.1176	0.68599	9.500	0.000
23	CM 3	4.0588	0.64860	9.519	0.000
24	SCM 2	3.9412	1.22947	4.464	0.000
25	CM 5	3.9412	1.12657	4.871	0.000
26	MM 1	3.8824	0.68599	7.500	0.000
27	SCM 1	3.6471	0.98110	3.846	0.001
28	SCM 8	3.5882	1.35104	2.539	0.016
29	SCM 5	3.5882	1.10420	3.106	0.004
30	MM 6	3.5294	1.05127	2.936	0.006
31	SM 3	3.5294	0.78760	3.919	0.000
32	CM 6	3.2353	1.12973	1.214	0.233
33	HRM 5	2.8824	1.14851	−0.597	0.554

Note Please refer to Table 4.5 for the definitions of the abbreviations in this table

(Eiadat 2008). It is also gathered that commitment from the subcontractors can be anchored for through contractual means, so that they are compelled to and responsible for meeting the specific CONQUAS score as well (Saraph et al. 1989).

On the other hand, "sample testing of materials through an ITA to check for proper usage of materials (MM6)", "cross-checking by another party (CM6)" as well as "supervision by the client or architect to monitor the CONQUAS management workflow of the main contractor (HRM5)" are the three least important factors in influencing the CONQUAS score. It has been inferred from the three interviews that cross-checking is not considered as important because much time and manpower is required and contractors feel that it is not worthwhile to do so and would rather focus on other aspects of the project. This was further explained by Low and Teo (2004) who found that contractors are already competing with time to complete projects, and hence, even if a CONQUAS management framework is in place, there is no guarantee that they will actually be implemented. Therefore, only those factors that are evidently effective will be considered as important and retained over time (Koh and Low 2008).

Furthermore, the main contractors may have their own system of handling the CONQUAS workflow and do not like any form of supervision by another party. This was explained by Ms B who said that:

> The client or architect can only supervise or monitor but the main CONQUAS management workflow is usually being controlled by the main contractor, hence the low mean importance rating obtained.

In the same way, it is not considered important for the client or architects to supervise as pointed out by Mr C who felt that:

> Some developers may not be that concerned about CONQUAS and totally allocate such risks and responsibilities to the contractors.

It follows that the relatively low importance rating of using the service of an ITA may be because ITA is primarily employed for the reason of meeting the contractual requirements. This was understood from Ms B who explained that:

> ITA is also not implemented for the reason of ensuring high CONQUAS score but achieving high score is a consequence of that.

Next, the one-sample t test is used to compare the mean score of the responses received to a "neutral" test value of 3 and also to show that the normal distribution assumption is fulfilled based on the central limit theorem, having a sample size of more than 30.

- Null hypothesis, $H_{1.0}$: $\mu = 3 \rightarrow$ It indicates neutrality of the importance of the CSFs in influencing the CONQUAS scores.
- Alternate hypothesis, $H_{1.1}$: $\mu \neq 3 \rightarrow$ It indicates non-neutrality of the importance of the CSFs in influencing the CONQUAS scores.

The results are interpreted by comparing the significance test level, p, against the level of significance of 0.05 such that

- If $p > 0.05$, $H_{1.0}$ will not be rejected which means that the importance level of the CSFs is not significant in influencing the CONQUAS scores.
- If $p \leq 0.05$, $H_{1.0}$ will be rejected, but this does not mean that the importance level of the CSFs is significant in affecting the CONQUAS scores. A second step of comparing the t-statistics against the critical value is required. The critical value of t at $p = 0.05$ for 37 degree of freedom $(n - 1)$ is 2.026. If $t > 2.026$, it means that the importance level of the CSFs is significant in influencing the CONQUAS scores. However, if $t < -2.026$, it means that the importance level of the CSFs is significantly inapplicable to having an effect on the CONQUAS scores.

As a whole, the importance level of 31 CSFs variables is statistically significant to have some bearing on the CONQUAS scores at the 95 % confidence interval ($p \leq 0.05$), while the remaining two CSFs variables which are not statistically significant have to be rejected. Hence, the null hypothesis $H_{1.0}$ would be rejected, indicating the non-neutrality of the importance level of 31 CSFs in influencing the CONQUAS scores. This also supports the notion made earlier that these CSFs could be distinguished into the "Features" and "Basic Attributes" groups. The results also suggest that the first 23 CSFs variables which are statistically significant and have a mean rating above 4, signifying relatively higher level of importance to influence the CONQUAS scores, will be classified under the "Features" group, while the other CSFs variables will remain as "Basic Attributes", indicating minimum level of importance. Therefore, it is inferred that contractors should pay more attention to these 23 CSFs in the "Features" group, having a higher level of importance and playing a more critical role in contributing to the achievement of high CONQUAS scores.

By cross-referencing Mr A's comments below with the CSFs tested in the survey, it is noted that those factors mentioned, although named differently, meant the same thing and it appears that they are all in the "Features" group. This is evident as Mr A said:

> Most of the time, the influencing factors for achieving high CONQUAS scores are management support, awareness on the importance of achieving the target, site staff awareness and understanding of the scoring system and penalties, regular site inspection and close monitoring of progress and scores.

However, both Mr A and Ms B pointed out several factors not tested in this statistical analysis which are important as well. The exclusion of such factors may in fact be important given that Ms B's company have only completed three CONQUAS projects and that they tend to avoid projects which are more stringent in the workmanship quality requirement as they are not confident of doing so which led to their poor CONQUAS score record. This implies that the experience and expertise of the QMU is important to yield a good CONQUAS score. This is evident as Mr A recommended the consideration of:

> Other external factors such as the experience of the site team in managing CONQUAS, the discipline of the people involved and maybe the requirements of the projects.

It is also believed that many quality problems can be reduced if greater attention is paid to the quality of communication. An effective quality effort will require the participation of everybody in the QMU, and clear and precise communication with other stakeholders is also important to achieving workmanship quality excellence (Wang 1998). This is evident as Ms B commented that:

> To a certain extent, having a good relationship with the BCA assessors can be beneficial too. All that said, theoretically, BCA assessors are supposed to monitor the work qualities as a third party which has no grounds on either side in order for their reports to be fair… Communication to the workers regarding the proper usage of materials, especially for new products, is important as no matter how good the material may be, it may not perform as expected if it is being handled, stored or used in the wrong way.

Moreover, the achievement of high CONQUAS score is constrained by clients who are still focused on cost and time rather than on quality. This is evident as Poon and Xu (1997) found that organisations needed to minimise costs due to competition based on the lowest price. As a result, staffs were often overloaded with work, and hence, they were unable to handle CONQUAS management issues effectively. This implies that clients and contractors are unwilling to support the achievement of high CONQUAS scores and simply do just enough to meet the minimum requirements without caring too much for the quality spirit.

Overall, these factors which are not included in the survey could be a drawback to the rigour of the survey analysis. But even so, it is believed that the impact on the results of the survey is minor as it was acknowledged by the three interviewees that implementation of all 33 CSFs would be more than sufficient to attain a high CONQUAS score. In particular, Mr A said that:

> If a company is able to achieve all these 33 CSFs successfully, I think that achieving a score above 90 is not a problem.

6.3.1 Mean Importance Rating Based on the Five Categories

Based on the category that each of the 33 CSFs belongs to, the mean importance rating is computed and HRM is found to be the most important category to influence the CONQUAS score as shown in Fig. 6.3. This outcome was explained by Mr C who commented that:

> The factors are rather comprehensive and the factors under human resource management are the pre-requisites to meet the factors under other aspects like materials and construction management.

All three interviewees noted that the main approach to managing CONQUAS would certainly involve factors under HRM. HRM is the process of ensuring that there is a competent and adequate team of people to develop strategies to achieve high CONQUAS scores. A firm's human resources offer it the opportunity to create a competitive advantage if the firm chooses to do so (Hamel and Prahalad 1994). Therefore, the decisions that contractor makes for its HRM strategies are important and should not be made by default (Maloney 1997). It should be thought through carefully as every construction project is unique, consisting of different teams of people, and hence, the HRM approach may have to be tweaked accordingly so as to be able to handle different CONQUAS processes effectively and efficiently. Mr A also mentioned that:

> Our approach is more towards working harder on human resource management…

In construction, considerable resources are being allocated to planning as studies have shown the effectiveness of planning techniques to the achievement of project goals (Laufer and Tucker 1987). By applying this concept, it is noted that quality would suffer if the CONQUAS planning process was not executed properly by the QMU. This can be seen in Mr A's company with a good track record of CONQUAS scores. Mr A mentioned that their focus to managing CONQUAS projects is on:

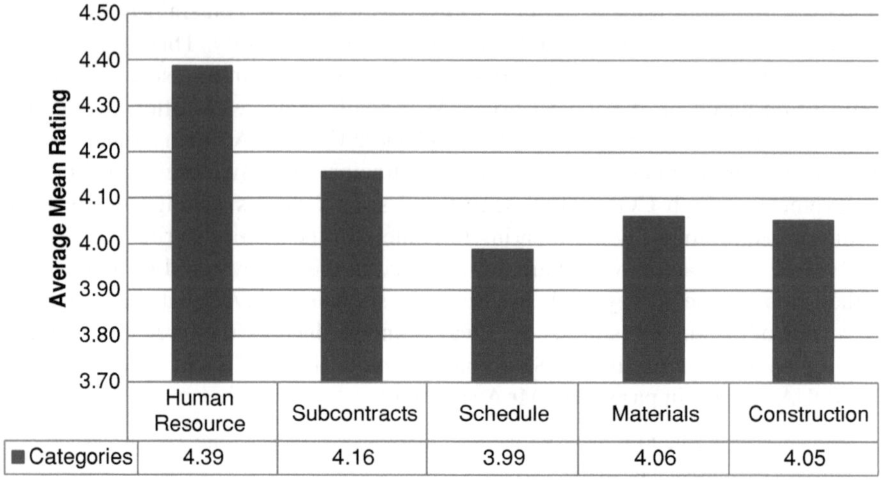

Fig. 6.3 Average mean importance level in terms of the five categories

> Planning and ensuring that what is being planned can be realised on site. …and at the same time will ensure that other factors affecting the CONQUAS scores are not neglected.

Similarly, the focus of Ms B's company with a poor track record of CONQUAS scores to achieve high CONQUAS score is on HRM as well. As Ms B explained:

> The QMU acts by setting out specific broad goals for the 'ultimate targets of quality'…to meet the target score identified in our company's quality policy. The CONQUAS committee comprises mostly of experienced site personnel who understands the site conditions to ensure workmanship quality. Therefore, we are able to set up benchmarks that are practical and achievable on site.

Moreover, Ms B's company have to rely on veterans' expertise to advance their CONQUAS performance. As stated by Ms B:

> An experienced senior who has been involved in many CONQUAS projects…He has given us a lot of tips in preventing defects from occurring by pre-empting us beforehand. He will also warn us to take extra care on areas, based on his experience, which have a high probability of non-compliance.

In today's highly competitive environment, managing people effectively can have a significant impact on the results (Hubbard et al. 1990). A study by Culp and Smith (1992) revealed that a well-trained project manager is a key factor linked with project success because as a team builder, he or she can create an effective team. Hence, by ensuring that adequate training is provided to the staffs will certainly help in managing CONQUAS more effectively and efficiently (Rao et al. 1987). Likewise, Mr C emphasised that:

> Staff training on CONQUAS is very important…I think that it is a basic requirement for all staffs who will be involved in duties related to workmanship quality. The latest CONQUAS edition has also provided bonus points for projects which employ certified CONQUAS supervisor or manager.

6.4 Usage Level of the Critical Success Factors

Even though contractors realised the importance of these CSFs, it is unknown whether they have really implemented them in their projects. In other words, even though contractors are able to acknowledge that "Features" are more critical than "Basic Attributes", it does not necessarily mean that contractors will implement them and place "Features" as their priority over "Basic Attributes". Hence, the main purpose of the third section of the survey questionnaire is to identify the "Features" which are not commonly used by contractors in their CONQUAS management workflow.

Similarly, the one-sample t test is used to compare the mean score of the responses received to a "neutral" test value of 0.5 and also to show that the normal distribution assumption is fulfilled based on the central limit theorem, having a sample size of more than 30.

- Null hypothesis, $H_{2.0}$: $\mu = 0.5 \rightarrow$ It indicates neutrality of the extent of adoption of the CSFs in the CONQUAS management workflow to the achievement of high CONQUAS scores.
- Alternate hypothesis, $H_{2.1}$: $\mu \neq 0.5 \rightarrow$ It indicates non-neutrality of the extent of adoption of the CSFs in the CONQUAS management workflow to the achievement of high CONQUAS scores.

The results are interpreted by comparing the significance test level, p, against the level of significance of 0.05 such that

- If $p > 0.05$, $H_{2.0}$ will not be rejected which means that the extent of adoption of the CSFs is not significant in the CONQUAS management framework to the achievement of high CONQUAS scores.
- If $p \leq 0.05$, $H_{2.0}$ will be rejected, but this does not mean that the usage level of the CSFs is significant in the CONQUAS management framework to the achievement of high CONQUAS scores. A second step of comparing the t-statistics against the critical value is required. The critical value of t at $p = 0.05$ for 37 degree of freedom ($n - 1$) is 2.026. If $t > 2.026$, it means that the extent of adoption of the CSFs is significant in the CONQUAS management framework to the achievement of high CONQUAS scores. However, if $t < -2.026$, it means that the usage level of the CSFs is significantly inapplicable to the CONQUAS management framework to the achievement of high CONQUAS scores.

The results obtained are presented in Table 6.4 and ranked accordingly from the highest mean usage rating to the lowest mean usage rating.

Based on the above understanding, it can be observed from Table 6.4 that the extent of adoption of 29 CSFs variables is statistically significant ($p \leq 0.05$, $t > 2.026$) and four CSFs (indicated with an asterisk) are not statistically significant in the CONQUAS management workflow to the achievement of high CONQUAS scores. Hence, the null hypothesis $H_{2.0}$ would be rejected and this shows the non-neutrality of the usage level of the CSFs to the achievement of high CONQUAS scores. This supports the alternate hypothesis $H_{2.1}$ that the higher the adoption of the

Table 6.4 Usage rating of the 33 CSFs

Rank	Critical success factors	Mean usage rating	Standard deviation	*t* value	Significance (two-tailed)
1	SCM3	0.7941	0.41043	4.179	0.000
2	CM7	0.7647	0.43056	3.585	0.001
3	HRM6	0.7059	0.46250	2.596	0.014
4	SCM9	0.7059	0.46250	2.596	0.014
5	HRM1	0.6471	0.48507	7.778	0.000
6	HRM3	0.6471	0.48507	7.778	0.000
7	SCM4	0.6471	0.48507	7.778	0.000
8	SCM7	0.6471	0.48507	7.778	0.000
9	SM1	0.6471	0.48507	7.778	0.000
10	SM2	0.6471	0.48507	7.778	0.000
11	MM2	0.6471	0.48507	7.778	0.000
12	MM3	0.5882	0.49955	5.745	0.000
13	CM1	0.5882	0.49955	5.745	0.000
14	SCM1	0.5588	0.50399	6.866	0.000
15	SM4	0.5588	0.50399	6.465	0.000
16	SCM5	0.5294	0.50664	8.899	0.000
17	CM4	0.5294	0.50664	8.899	0.000
18	HRM2	0.5000	0.50752	5.745	0.000
19	HRM4	0.5000	0.50752	5.745	0.000
20	SCM2	0.5000	0.50752	6.093	0.000
21	SCM6	0.5000	0.50752	5.745	0.000
22	HRM7	0.4706	0.50664	6.465	0.000
23	MM5	0.4706	0.50664	8.899	0.000
24	CM5	0.4412	0.50399	5.104	0.000
25	CM2	0.4118	0.49955	4.806	0.000
26	MM4	0.3824	0.49327	5.104	0.000
27	SM3*	0.3235	0.47486	−2.167	0.038
28	CM3*	0.2941	0.46250	−2.596	0.014
29	SCM8*	0.2353	0.43056	−3.585	0.001
30	MM6*	0.2353	0.43056	−3.585	0.001
31	MM1	0.2059	0.41043	2.925	0.006
32	CM6	0.1471	0.35949	2.385	0.023
33	HRM5	0.1176	0.32703	2.098	0.044

Note Please refer to Table 4.5 for the definitions of the abbreviations in this table;
Asterisks indicate factors which are not statistically significant in the CONQUAS management workflow to the achievement of high CONQUAS scores

CSFs, the greater the resultant CONQUAS score will be than when the CSFs are not adopted, allowing certain firms to achieve higher CONQUAS scores than the rest.

Therefore, it can be ascertained that out of the 23 "Features", those with a low mean rating below 0.5 will certainly require more attention and have to be properly implemented on a full scale in order to achieve high CONQUAS scores. These are "setting a benchmark to gauge and control the CONQUAS performance (HRM7)", "protection of materials after completion of that portion of works (MM5)", "constructing mock-ups to check for implications with other trade of works (CM2)", "proper materials handling and storage (MM4)" as well as "field demonstration by labourers to showcase their understanding of the workmanship quality required (CM3)". This suggests that contractors knowingly ignore the criticality of these five "Features" as pointed out by Mr A who acknowledged that the 33 CSFs are recognised industry-wide by all stakeholders in the project:

> Many a times, main contractors, sub-contractors and site staff do understand the importance, but when it boils down to practicing what they preach, it is a totally different mindset. So the policing part comes into play and how the QA guys do their inspection and how well the sub-contractors cooperate.

Therefore, contractors should now work harder on these five CSFs in their CONQUAS management workflow to improve their workmanship quality and attain higher CONQUAS scores. To do so, it is believed that the BCA could recommend a CONQUAS management framework and that corporate policy should also lay out clearly the CONQUAS management process. Such national and corporate policies are to be implemented on a full scale to propel the entire industry to strive to achieve high CONQUAS scores. They are essential to act as a guiding principle and mechanism for ensuring that the quality performance is able to meet requirements and in the best interests of the company too (Eiadat 2008). In conclusion, it appears that the construction of a CONQUAS management framework can also illustrate the importance of getting the commitment of the industry and individual companies to achieving high CONQUAS scores.

6.4.1 Mean Usage Rating Based on the Five Categories

Based on the category that each of the 33 CSFs belongs to, the mean usage rating is computed and the usage level of all five categories are about the same as shown in Fig. 6.4. The slightly higher mean usage rating of the SCM category may be due to the fact that subcontractors are the ones who execute the construction work on site, and hence, they are the people who are directly responsible for the poor workmanship quality. Subcontractors are the ones working in the field to deliver the final building product and have to be answerable for their actions. Therefore, the main contractors often have to tie them down contractually so that they will have to be made partly accountable for the CONQUAS score performance as well.

6.5 Correlation Analysis

6.5.1 Coefficient of Determination

Furthermore, to test whether the extent of adoption of the CSFs is dependent on whether it is important or not, the value of R^2 has to be found. R^2 measures the percentage of variation in the y variable (usage rating) which can be attributed to

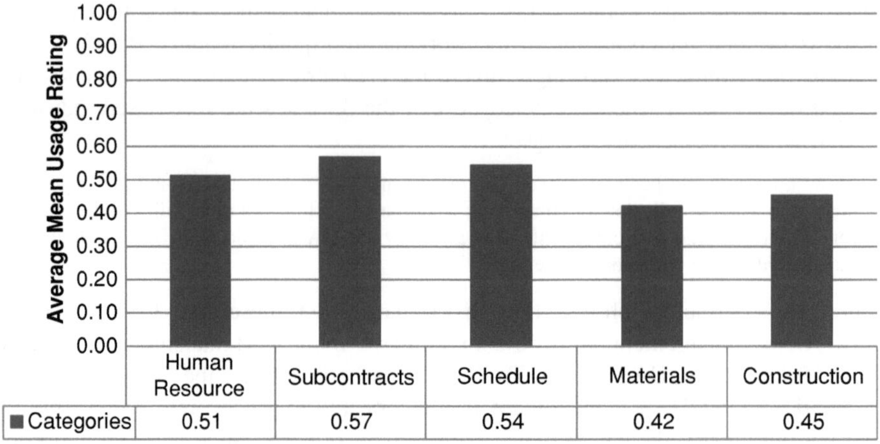

Fig. 6.4 Average mean usage rating in terms of the five categories. *Note* The value of 1.00 means total usage of all the factors under the respective category; the value of 0.00 means absolute non-usage of all the factors under the respective category

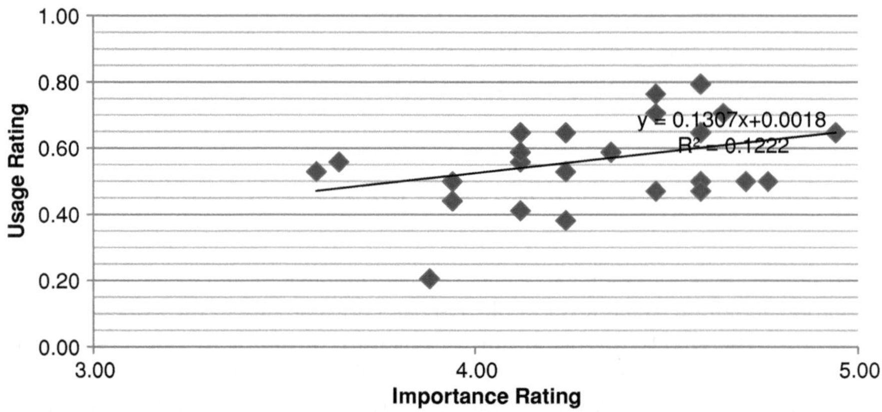

Fig. 6.5 Usage rating against importance rating

variation in the *x* variable (importance rating). Here, only the 27 CSFs which were not rejected in the mean importance and usage-level significance test were taken into account to validate the relationship as the results of the other six CSFs only occurred due to chance. As seen in Fig. 6.5, the mean importance of the 27 CSFs with its mean usage is poorly correlated where the correlation coefficient squared is only 0.1222. This implies that about 12 % of the variation in usage rating data is due to the variation in the importance rating data. The results appear to suggest that the importance rating is affecting the usage rating to a very small extent.

It was initially hypothesised in Chap. 1 that the perceived importance of the CSF does not lead to the actual adoption and practice of the CSF. In other words, it was hypothesised that contractors were not likely to implement the CSF even if they perceived it as being important. This seems to be consistent with the results obtained which showed that the contractors' perception of higher importance of the CSF does not lead to the increased adoption, as there is an extremely poor coefficient of determination. Hence, the null hypothesis $H_{3.0}$ would be accepted which suggests that the extent of adoption of the CSFs is very much independent of its importance level.

The poor R^2 value can also be explained by the fact that the level of usage of each of the CSFs varies between different firms, and this was elaborated by Ms B who felt that:

> It would be great if contractors are able to implement all these 33 factors, but normally, such high mean importance rating factors tend to be much harder to achieve, in terms of having to spend more time and manpower in order to achieve. In that case, contractors may choose to spend their time and manpower on achieving more factors that have lower mean importance rating than just to spend all their time and manpower on a single high mean importance rating factor that at the end of the day, may still not be achievable despite all the time and manpower spent.

Nonetheless, there may still exist some "Features" (more important factors) which are easy to achieve as the effort required to achieving high CONQUAS scores is just proper planning which can be built up from their experiences in managing CONQUAS projects as highlighted by Ms B. It is also noted that contractors should not remain status quo but should strive to improve their CONQUAS management activities to continue to achieve higher CONQUAS scores even with the constant development made to the CONQUAS scheme by the BCA.

In addition, despite the fact that a substantial number of respondents come from reputable firms and have vast experience in managing CONQUAS, they may have a misconception for which factor is more important to use than the rest. Moreover, some factors may only be applicable in certain situations too. In particular, a relatively low ranking of both the importance level and usage level of the "adequacy of sample size to be prepared for inspection (SCM5)" is contrary to Mr A's viewpoint:

> Although the minimum number of sample to be prepared is as indicated in the CONQUAS standards, based on my experience, the CONQUAS scores tend to be better if the entire block is ready for inspection. But in reality, this may be hard to achieve due to schedule issues and it is common that at most, only half of the block is ready and this may leave a poor impression to the assessors in general.

Although there may be a possibility of lurking variables such as insufficient time and manpower to implement the respective CSFs which may lead to non-usage of the CSFs despite perception of high importance, it is believed to be insignificant in affecting the results of this study as supported by both Mr A and Ms B who expressed that it is predictable that a poor coefficient of determination (0.1222) would be obtained as they were absolutely certain that the more important factor does not necessarily equate to higher degree of usage and vice versa. Mr A felt that:

> I don't think that the higher mean importance rating means that contractors will input them as part of their strategies to achieve high CONQUAS scores.

Similarly, Ms B also said that:

> Normally, contractors have a target CONQUAS score in mind and will then decide on the factors that are more achievable within their available resources, of which, most of the time will not be 0the high mean importance rating factors unless it is their forte.

Table 6.5 Statistically significant positive correlation of importance rating

Pearson correlation coefficient (r)	Coefficient general interpretation	No. of significant pairs of positive coefficient
≥0.70 to ≤1.00	Strong	20
≥0.40 to <0.70	Moderate	155
≥0.00 to <0.40	Weak	69
Total positively significant pairs		244

6.5.2 Pearson Correlation

The Pearson correlation coefficient, r, is utilised to examine the relationship between each CSFs with every other CSFs. Two sets of correlations were computed to analyse the relationships between variables, that is, the respondents' perceived importance of each CSF with every other CSF as well as the actual adoption and practice (usage level) of each CSF with every other CSF. The results of the analysis would be used to discuss the relationship and strength of association between the variables tested. This study will follow the reference point used by Reed and Frankham (2003) where $r \geq 0.70$ means strong correlation, $r < 0.40$ means weak correlation and the values in between will mean moderate correlation. The sections below will now highlight on the end results to support the acceptance or rejection of Hypotheses 4 and 5 presented in Chap. 1.

- Null hypothesis, $H_{4.0/5.0}$: $r = 0 \rightarrow$ Absence of correlation relationship
- Alternate hypothesis, $H_{4.1/5.1}$: $r \neq 0 \rightarrow$ Presence of correlation relationship.

6.5.2.1 Correlation Between Importance of Each CSF and Every Other CSF

This section aims to find out the correlation between the respondents' perceived importance level of each CSF with every other CSF. Out of 528 pairs of variables tested, 246 pairs gave a statistically significant correlation ($p \leq 0.05$; df = 526; $r > |0.088|$), with 244 pairs having significant positive correlation as indicated in Table 6.5. The detailed results can be found in Appendix E. This means that about 46.2 % of them are interrelated, where higher importance of the particular CSF in the CONQUAS management framework will lead to higher importance of the respective CSF.

The results suggest that there is a relatively strong and positive correlation for twenty pairs of variables, while 69 pairs gave a relatively weak and positive correlation. Generally, CSFs under the same category gave a higher correlation which is justifiable since they are targeting similar areas of concern in the CONQUAS management framework. Moreover, HRM CSFs have a higher correlation with SCM CSFs and this is because people are the ones who will be managing the subcontractors. This link between the subcontractors and the QMU of the main contractors has also been established in Mr A's CONQUAS management strategy which

Table 6.6 Statistically significant negative correlation of importance rating

No.	Factor 1	Factor 2	Pearson correlation coefficient (*r*)	Sig. (two-tailed)
1	HRM2	SCM8	−0.362	0.025
2	HRM4	SM3	−0.455	0.004

Note Please refer to Table 4.5 for the definitions of the abbreviations in this table

has proven to be effective, leading to an improvement in their CONQUAS track record. As Mr A explained:

> It is also important that this QA or QC site team consists of personnel who are responsible for fine tuning or touching up to prepare the area prior to inspection. Therefore, besides the inclusion of personnel from our company, we will also include the site supervisor of each subcontractor's firm in this QMU.

Furthermore, MM CSFs have a higher correlation with CM CSFs and this is because both categories are dependent on one another as the selection of building materials and construction methods will have to be discussed to find the most suitable and practical way to carry out the construction works (Anink et al. 1996). Therefore, this has reinforced the need for contractors to understand that the different CSFs have to work together as they are vastly interrelated and any one of them should not be overlooked as it will affect the eventual CONQUAS management workflow and may also hinder activities from progressing smoothly.

On the other hand, Table 6.6 shows the two pairs of variables which gave a statistically significant, moderately weak negative correlation with regard to the importance responses. This means that a higher importance of the particular CSF will lead to a lower importance of the respective CSF. This may be due to the fact that when top management are truly committed to advance the CONQUAS management performance (HRM2), they need not resort to imposing penalties on the subcontractors if they do not meet the targeted CONQUAS score stipulated in the contract (SCM8). In other words, contractors feel that as long as the QMU, which consists of different stakeholders such as the main contractor, subcontractors and the client, has the full commitment to handle the CONQUAS workflow, they do not have to make use of monetary penalties to force subcontractors to perform and achieve good workmanship quality.

Moreover, when staffs are well trained and understand the CONQUAS requirements (HRM4), they may have acquired other means of achieving workmanship quality and need not arrange for phased construction to reap the learning curve on workmanship quality (SM3). This negative correlation suggests that contractors acknowledge that phased construction is not the only means of achieving workmanship quality and frequent trainings are important in order to keep up with changes and trends in the construction industry so that contractors will be aware of the improvements required to execute the CONQUAS management framework effectively to attain good CONQUAS scores. This accords well with Ms B's view who mentioned that:

> Arranging for phased construction to reap learning curve on workmanship quality is only workable if there is repetition in the buildings, that is, every storey is the same so that as one goes up, one learn from past mistakes and do better.

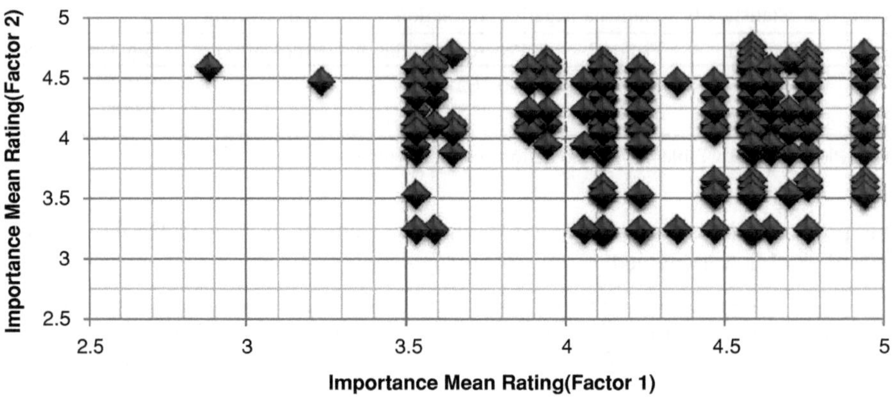

Fig. 6.6 Scatter plot diagram of the importance relationship between statistically significant pairs of CSFs

Table 6.7 Statistically significant positive correlation of usage rating

Pearson correlation coefficient (r)	Coefficient general interpretation	No. of significant pairs of positive coefficient
≥0.70 to ≤1.00	Strong	57
≥0.40 to <0.70	Moderate	204
≥0.00 to <0.40	Weak	42
Total positively significant pairs		303

Figure 6.6 shows the scatter plot diagram of the above 246 statistically significant pairs of variables and notes that the linear relationship is moderate since their average correlation is 0.486. But generally, it is observed that the relationship is positive and knowing the importance level of the particular CSF can help to infer something about the degree of importance of another CSF moderately. Therefore, even though only 47 % of the variables are statistically significant, it is determined that the null hypothesis $H_{4.0}$ set in Chap. 1 would still have to be rejected, meaning that the importance level of each of the CSFs is correlated with the importance level of other CSFs, to a reasonable extent.

6.5.2.2 Correlation Between the Usage Level of Each CSF with Every Other CSF

This section seeks to investigate the correlation and impact of the usage level of each CSF with every other CSF. The results showed that out of 528 pairs of variables tested, 305 pairs gave a statistically significant correlation ($p \leq 0.05$; df = 526; $r > |0.088|$), with 303 pairs having significant positive correlation as shown in Table 6.7. This means that about 57.4 % of them are interrelated, where a higher usage level of the particular CSF in the CONQUAS management

Table 6.8 Statistically significant negative correlation of usage rating

No.	Factor 1	Factor 2	Pearson correlation coefficient (r)	Sig. (two-tailed)
1	SCM8	SM4	−0.365	0.024
2	SM4	MM1	−0.365	0.024

Note Please refer to Table 4.5 for the definitions of the abbreviations in this table

framework will lead to a higher adoption of the respective CSF as well. The detailed results can be found in Appendix F.

It is noted that the closer is the value of r to 1, the greater is the degree of linear statistical relationship between the two variables. This means that with more practice of the particular CSF, there will be a greater increase in the usage of the respective CSF. Of which, 29 of them are perfectly positively correlated ($r = +1$). This is surprising as perfect correlation is rare. This could be because some factors may intersect between two categories and hence the extremely high correlation. For example, formation of the QMU (HRM1) and collaborative efforts between the subcontractors and main contractor (SCM4) may be viewed as similar and dependent on each other since subcontractors are also part of the QMU.

A close observation of the data did not result in a noticeable trend which might be due to extraneous factors that mask the true relationship between them. For example, only 38 respondents voluntarily replied to the survey conducted and those who did not reply may have vastly different opinions on the CSFs, and hence, even though this study has satisfied the criteria to assume that the variables are sampled from a Gaussian population to do this Pearson statistical analysis, it may still not be entirely effective to determine results that can be generalised as a whole.

Next, the results in Table 6.8 show that the extent of adoption of two pairs of variables is negatively and weakly correlated. It can be interpreted from the results that a higher adoption of imposing penalties on the subcontractors when they do not meet the targeted CONQUAS score (SCM8) will lead to decreased adoption of booking assessment schedule to tie in with the site progress (SM4). Further, a higher practice of choosing materials thoroughly (MM1) will also lead to decreased practice of booking assessment schedule to tie in with the site progress (SM4). This may also be attributable to the late arrangements made by the contractors which lead to difficulty in fixing an appointment time when the BCA assessors are available. This is coupled by the fact that there are many CONQUAS projects in Singapore but only a small number of BCA assessors available to attend to concurrently on top of their other scope of works such as holding CONQUAS trainings for stakeholders involved, reviewing the CONQUAS scheme and collating defects trend observed. As highlighted by Ms B,

> There are only a handful of assessors, maybe 15…

Following that, Fig. 6.7 shows the scatter plot diagram of the above 305 statistically significant pairs of variables. It is noted that the linear relationship is moderate given that the average correlation is only 0.532. But generally, it can be

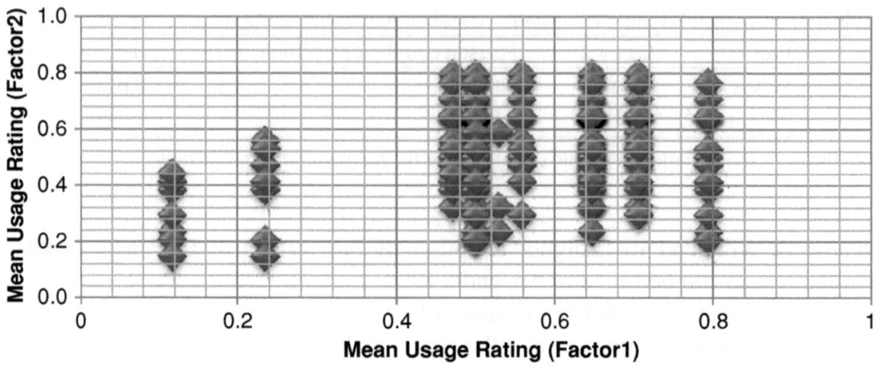

Fig. 6.7 Scatter plot diagram of the usage relationship between statistically significant pairs of CSFs

seen that the relationship is positive, and hence, contractors will be able to visually identify relationships between the first and the second entries of paired data moderately. Therefore, since there is statistical significance for the correlation of slightly more than half (58 %) of the pairs of variables tested, the null hypothesis $H_{5.0}$ formulated in Chap. 1 would be rejected following the results of the correlation, meaning that the extent of adoption of each of the CSFs is correlated with the extent of adoption of the other CSFs, to a reasonable extent.

6.5.3 *Concluding Statement*

In conclusion, the importance level of each of the CSFs is positively correlated with the importance level of the other respective CSFs most of the times. Similarly, the usage level of each of the CSFs is positively correlated with the usage level of the other respective CSFs in general. This outcome is consistent with the research's objective of implementing a CONQUAS management framework by associating all 33 CSFs together and guiding contractors on the adoption in actual practice to enhance workmanship quality. It should be reiterated that the correlation does not imply that one CSF causes the other CSF to occur.

6.6 Linear Regression

Although correlation is a powerful method to describe and test associations between continuous variables, it is not able to predict the value of one dependent variable, from measurements of the other independent variable. Hence, to quantify how the usage of a particular CSF affects the CONQUAS score, a regression model was developed. The stepwise regression selection procedure was chosen to reduce

Table 6.9 R^2 and adjusted R^2

Model summary				
Model	R	R^2	Adjusted R^2	Std. error of the estimate
1	0.968	0.937	0.827	2.289

Table 6.10 Variables removed

Variables removed	
1	HRM2—Commitment of top management
2	HRM3—Supportive and cooperative employees
3	HRM5—Supervision by client or architect
4	SCM3—Skill level of labours
5	SCM4—Collaborative efforts between subcontractors and main contractor
6	SCM5—Adequacy of contract period and contract sum
7	SM3—Arrange for phased construction
8	MM3—Inspect materials upon delivery
9	MM6—Sample testing through ITA
10	CM1—Review of shop drawings
11	CM2—Construct mock-ups
12	CM6—Cross-checking

Criteria: Probability of F to enter ≤ 0.050; probability of F to remove ≥ 0.100

the number of variables (CSFs) so that the resulting regression equation is easy to understand, interpret and work with, consisting of only the most significant variables to effectively predict the CONQUAS score. Generally, one could continue to add predictors (CSFs) to the model which would continue to improve the ability of the predictors to explain the dependent variable (average CONQUAS score obtained in the recent three years), although some of this increase in R^2 would be simply due to chance variation in that particular sample. Here, it is noted that the original R^2 value without removing outliers is 0.937, an extremely high value which may be due to presence of relationships that are not statistically significant as identified in the above section. The best model summary is tabulated in Table 6.9.

This best model consisting of only statistically significant variables computed that twelve CSFs (Table 6.10) have to be eliminated as they are not normal, having relationships that are collinear. With that, the adjusted R^2 value became 0.827, signifying that the remaining 21 CSFs can account for about 83 % of the variability in the CONQUAS score. It is noted that the remaining 17 % may be due to the demographic profile of the respondents and their company and the discipline of the people managing the project as suggested by the interviewees as well. This means that there are actually other variables which may exert influence over the CONQUAS score, some of which are unpredictable or difficult to gather the data.

Moreover, the F test which is used to compute the significance of each added variable via stepwise to the explanation reflected in R^2 shows that the regression model is significant ($p \leq 0.05$) as presented in Table 6.11. This implies that the regression equation does have some validity in fitting the data. The independent variables (CSFs) are not purely random with respect to the dependent variable (CONQUAS score).

Table 6.11 *F* statistics

ANOVA						
Model		Sum of squares	df	Mean square	*F*	Sig.
1	Regression	936.173	21	44.580	8.507	0.000
	Residual	62.886	12	5.241		
	Total	999.059	33			

Table 6.12 Regression equation coefficients

Model		Unstandardised coefficients	Standardised coefficients	Sig.
		B	Beta	
1	(Constant)	75.117		0.000
	HRM1	1.014	0.093	0.001
	HRM4	1.164	0.098	0.009
	HRM6	0.429	0.039	0.008
	HRM7	1.197	0.110	0.003
	SCM1	−0.117	−0.009	0.001
	SCM2	1.096	0.101	0.040
	SCM6	0.276	0.023	0.000
	SCM7	−0.134	−0.012	0.007
	SCM8	−2.461	−0.227	0.025
	SCM9	0.468	0.039	0.005
	SM1	1.864	0.169	0.002
	SM2	0.262	0.024	0.025
	SM4	0.181	0.017	0.008
	MM1	3.064	0.229	0.025
	MM2	−7.314	−0.645	0.008
	MM4	1.817	0.166	0.025
	MM5	1.063	0.098	0.001
	CM3	4.436	0.403	0.004
	CM4	4.432	0.331	0.019
	CM5	2.974	0.272	0.014
	CM7	1.271	0.117	0.004

Note Please refer to 4.5 for the definitions of the abbreviations in this table

The *t* statistics test showed that 21 CSFs are statistically significant ($p \leq 0.05$) as shown in Table 6.12.

Therefore, the final derived regression equation consisting of 21 CSFs is

Predicted CONQUAS score = 75.117 + 1.014 (HRM1) + 1.164 (HRM4) + 0.429 (HRM6) + 1.197(HRM7) **− 0.117 (SCM1)*** *+ 1.096 (SCM2) + 0.276 (SCM6)* **− 0.134 (SCM7)* − 2.461 (SCM8)*** *+ 0.468 (SCM9) + 1.864 (SM1) + 0.262 (SM2) + 0.181 (SM4) + 3.064 (MM1)* **− 7.314 (MM2)*** *+ 1.817 (MM4) + 1.063 (MM5) + 4.436 (CM3) + 4.432 (CM4) +2.974 (CM5) +1.271 (CM7)*

Note

1. Please refer to Table 4.5 for the definitions of the abbreviations in this equation.
2. To calculate the predicted CONQUAS score, for each factor in bracket, please insert the value "1" which means that the firm uses that particular factor or insert the value "0" which means that the firm do not use that particular factor when managing CONQUAS projects.
3. To reiterate, regression equation has to be based on an objective approach, and hence, the usage rating data were used instead of the importance rating data. Thus, the firm is deemed to be either utilising the factor (value "1") or not utilising the factor (value "0") only. Inserting intermediate values between 0 and 1 is not encouraged.
4. Asterisks indicate factors which have a negative coefficient.

This equation shows that the maximum possible CONQUAS that can be obtained is 93.4 (when all 21 CSFs are implemented), while the minimum possible is 72.4 (when all 21 CSFs are not used). The regression model shows that "controlling the number of subcontractors working on the project (SCM1)", "awarding incentives to subcontractors if their targeted score is met (SCM7)", "imposing a penalty to subcontractors if their targeted score is not achieved (SCM8)" and "select materials which can be better managed during construction (MM2)" have a negative coefficient, which means that the implementation of any of these four CSFs may potentially lower the CONQUAS score. It is believed that this drop in CONQUAS score is unusual and may only be applicable in situations where contractors themselves are not driven to meet these contractual conditions to attain high CONQUAS scores and simply seek to realise the basic requirements which can no longer meet the expectations and demands of the clients in today's rising standard of construction workmanship quality. Otherwise, this decrease may also be due to the fact that contractors do not have the expertise or a higher level of skill is required to achieve excellent workmanship quality in a particular scope of work.

On the other hand, "choose materials through a comparison process, considering their specifications and samples to be provided (MM1)", "field demonstration by labourers to showcase their understanding of the workmanship quality required (CM3)" and "conducting preparatory inspection using template checklist at every stage of work activity (CM4)" exert a larger positive effect as compared to the other CSFs, having a larger coefficient, and hence, contractors should experience a greater increase in CONQUAS score with the implementation of these three CSFs. Therefore, these three CSFs are important and should have a high degree of usage in the CONQUAS management framework to achieve high CONQUAS scores.

This model is useful as contractors would be able to estimate their CONQUAS scores, whether the requirements can be met, and yet within their limited resources and not sacrificing on other aspects such as productivity, buildability and constructability by deciding which CSFs to use or not to use and how it will impact the CONQUAS score. In other words, this model tells us which CSFs contractors should focus on in order to still achieve high CONQUAS scores.

Unlike Mr A and Mr C who are supportive and optimistic of the potential of this equation, Ms B was of the opinion that:

> This may only be practical to firms who lack the expertise in CONQUAS management. Big firms or those who have been performing well may feel that it is redundant since they may already have a well-established system in place and knows what to do in order to meet their target scores, although not quantified in exact numbers.

Furthermore, since these CSFs are correlated among themselves to a certain extent as determined in the Pearson correlation analysis, the common interpretation of the change in CONQUAS score when a particular CSF is not adopted while all other CSFs are held constant is not fully applicable. This is due to the fact that with highly correlated CSFs, it is difficult to attribute changes in the CONQUAS scores to one of the CSFs rather than another. Further, in perfect correlation, regression analysis breaks down completely, and the regression coefficient estimates will then have infinite variances. As the degree of collinearity among the CSFs increases, effects on the CONQUAS scores are much less significant. Hence, whatever the source of such multicollinearity, it is important to be aware of its existence as a limitation to the regression equation obtained.

6.7 Independent-Samples Test

The different importance rating and degree of usage of the CSFs may be due to limited resources as mentioned by Mr A and Ms B, especially the financial capability of the contractors, their category of financial grade. Mr C also mentioned that the deciding factor for utilising the CSFs is because of cost issues, and a very little percentage of the selection is based on the fact that better workmanship quality can be obtained.

This is substantiated as several publications concerning the implementation and application of similar practices (the 33 CSFs) have found that the majority of them are geared towards larger and reputable organisations with good track records (Goh and Ridgway 1994). This leads to the unfortunate conclusion by the smaller contractor's firms with limited financial resources that benefit such as attaining excellence in their workmanship quality and meeting high CONQUAS scores are out of their league. This may be coupled with their lack of experience and knowledge, the shortage of human resources, and time required for implementation (Haksever 1996; Hendricks 1992).

A study by Goh and Ridgway (1994) conducted among companies of a small scale also revealed that ISO 9000 certification was the endpoint in their quality drive. The same study also found that most companies did not see the advantage of analysing data related to cost of quality. On the other hand, companies of larger scale revealed that the reasons for adopting similar practices (the 33 CSFs) were promoting growth, changing customer expectations, making work more enjoyable and improving poor company performance (Shea and Gobeli 1995). This highlights the importance and adoption of the 33 CSFs for all financial categories of contractors so that they can improve their current quality management practices as well as quality of the built product, to achieve a good CONQUAS score.

Hence, the independent two-sample test is used to compare the responses received from A1 and A2 contractors to determine whether there is a distinction in their responses. B1 and B2 contractors were eliminated as the take-up rate of CONQUAS projects by them is significantly lower and with the low responses received from them, they are deemed unlikely to pose any substantial impact. Further, this is a small sample test which can only be used for two sample groups.

- Null hypothesis, $H_{6.0/7.0}$: $\mu_{A1} = \mu_{A2} \rightarrow$ The mean importance rating or extent of adoption of the CSFs of A1 and A2 contractors is not significantly different.
- Alternate hypothesis, $H_{7.1/7.1}$: $\mu_{A1} \neq \mu_{A2} \rightarrow$ The mean importance rating or extent of adoption of the CSFs of A1 and A2 contractors is significantly different.

Firstly, the significance test level will be compared against the significance value for Levene's test for equality of variances of 0.05 such that

- If the value is greater than 0.05, it suggests that the variance of the two groups is about the same and the "equal variance assumed" values would be used.
- If the value is equal or less than 0.05, it suggests that the variance of the two groups is significantly different from the other and the "equal variance not assumed" values would be used.

Secondly, the significance test level will be compared against the significance value for the *t* test of 0.05 such that

- If the value is greater than 0.05, the means of the two groups are not significantly different from each other.
- If the value is equal or less than 0.05, the means of the two groups are significantly different from each other.

6.7.1 Importance Rating

Table 6.13 shows that only twelve CSFs (indicated with a hash symbol) have higher mean importance rating by the A1 contractors as compared to the A2 contractors. This seems to be surprising and may be due to the fact that A1 contractors viewed the remaining 21 factors as something very basic and only rated those more important factors with a high rating. On the other hand, A2 contractors who are not as confident as the A1 contractors rated more factors with a high importance rating.

Next, out of 33 CSFs, only the means of two CSFs (indicated with an asterisk) of the two groups are significantly different as shown in Table 6.14. An analysis of Tables 6.13 and 6.14 did not reveal any observable trend to explain the results obtained. The only noteworthy point found was that there is a great divergence in

Table 6.13 Mean importance rating group statistics

		N	Mean	Std. deviation	Std. error mean
IHRM1#	A1	19	4.6316	0.68399	0.15692
	A2	8	4.5000	0.53452	0.18898
IHRM2#	A1	19	4.5789	0.50726	0.11637
	A2	8	4.2500	0.88641	0.31339
IHRM3	A1	19	4.4737	0.69669	0.15983
	A2	8	4.7500	0.46291	0.16366
IHRM4#	A1	19	5.0000	0.00000	0.00000
	A2	8	4.8750	0.35355	0.12500
IHRM5#	A1	19	3.3158	0.94591	0.21701
	A2	8	1.6250	0.74402	0.26305
IHRM6	A1	19	4.6842	0.47757	0.10956
	A2	8	4.8750	0.35355	0.12500
IHRM7#	A1	19	4.5263	0.51299	0.11769
	A2	8	4.1250	0.83452	0.29505
ISCM1	A1	19	3.6842	0.94591	0.21701
	A2	8	4.0000	1.06904	0.37796
ISCM2	A1	19	4.0526	1.26814	0.29093
	A2	8	4.1250	0.35355	0.12500
ISCM3	A1	19	4.5263	0.69669	0.15983
	A2	8	4.8750	0.35355	0.12500
ISCM4	A1	19	4.5789	0.76853	0.17631
	A2	8	4.7500	0.46291	0.16366
ISCM5#	A1	19	3.789	0.6306	0.1447
	A2	8	3.750	1.5811	0.5590
ISCM6	A1	19	4.6316	0.49559	0.11370
	A2	8	4.7500	0.46291	0.16366
ISCM7	A1	19	4.0526	0.77986	0.17891
	A2	8	4.2500	0.70711	0.25000
ISCM8	A1	19	3.4737	1.17229	0.26894
	A2	8	4.2500	1.48805	0.52610
ISCM9	A1	19	4.5263	0.69669	0.15983
	A2	8	4.8750	0.35355	0.12500
ISM1	A1	19	4.1579	1.01451	0.23275
	A2	8	4.3750	0.74402	0.26305
ISM2	A1	19	4.0526	0.97032	0.22261
	A2	8	4.3750	0.74402	0.26305
ISM3	A1	19	3.4737	0.84119	0.19298
	A2	8	3.6250	0.74402	0.26305
ISM4	A1	19	3.8947	0.93659	0.21487
	A2	8	4.5000	0.53452	0.18898
IMM1#	A1	19	3.9474	0.62126	0.14253
	A2	8	3.7500	0.88641	0.31339
IMM2#	A1	19	4.1579	0.76472	0.17544
	A2	8	4.1250	0.64087	0.22658
IMM3	A1	19	4.1053	0.73747	0.16919
	A2	8	4.1250	0.83452	0.29505

(Continued)

Table 6.13 (Continued)

		N	Mean	Std. deviation	Std. error mean
IMM4	A1	19	4.1579	0.76472	0.17544
	A2	8	4.2500	0.70711	0.25000
IMM5	A1	19	4.5263	0.69669	0.15983
	A2	8	4.7500	0.46291	0.16366
IMM6#	A1	19	3.6316	0.95513	0.21912
	A2	8	3.2500	1.48805	0.52610
ICM1	A1	19	4.3158	0.74927	0.17189
	A2	8	4.3750	0.91613	0.32390
ICM2#	A1	19	4.2105	0.78733	0.18063
	A2	8	4.1250	0.83452	0.29505
ICM3#	A1	19	4.0526	0.62126	0.14253
	A2	8	4.0000	0.75593	0.26726
ICM4#	A1	19	4.1579	0.68825	0.15789
	A2	8	4.0000	0.92582	0.32733
ICM5	A1	19	3.7368	1.09758	0.25180
	A2	8	4.1250	0.83452	0.29505
ICM6	A1	19	3.1579	1.21395	0.27850
	A2	8	3.5000	0.53452	0.18898
ICM7	A1	19	4.4211	0.69248	0.15887
	A2	8	4.6250	0.51755	0.18298

Note
1. The letter "I" in front of each of the factors refers to "Importance" of that respective factor
2. Hash symbols indicate factors rated with a higher mean importance rating by the A1 contractors as compared to the A2 contractors

the A1 contractors' and A2 contractors' views on getting the "supervision of the client or architect to monitor the CONQUAS management workflow of the main contractor (HRM5)". The reason may be because A2 contractors have not realised the benefits of getting the client or architect to supervise and hence do not think that they are important in influencing the CONQUAS score. On the other hand, the much higher mean score rated by the A1 contractors than the A2 contractors seems to suggest that supervision by the client or architect may be a significant determining factor in bringing about the higher CONQUAS scores achieved by the A1 contractors as compared to the A2 contractors. This is true to a certain extent in the sense that clients, as employers of the main contractor, by being engaged actively in the CONQUAS management workflow, will be able to spur contractors to perform the quality requirements accordingly so as to result in good workmanship quality and high CONQUAS scores.

As a whole, since the means of almost all the CSFs of the two groups are not significantly different from each other, the null hypothesis $H_{6.0}$ would not be rejected, indicating that in general, the importance rating by A1 and A2 contractors is about the same. Therefore, by and large, this accepts the null hypothesis $H_{6.0}$ that the different responses received by A1 and A2 contractors did not affect the responses of the importance level of the CSFs obtained.

Table 6.14 Independent-samples test for importance rating

		Levene's test for equality of variances		*t* test for equality of means						
		F	Sig.	*t*	df	Sig. (two-tailed)	Mean difference	Std. error difference	95 % confidence interval of the difference	
									Lower	Upper
IHRM1	Equal variances assumed#	0.092	0.764	0.484	25	0.633	0.13158	0.27211	−0.42884	0.69200
	Equal variances not assumed			0.536	16.862	0.599	0.13158	0.24564	−0.38699	0.65015
IHRM2	Equal variances assumed	8.692	0.007	1.226	25	0.232	0.32895	0.26830	−0.22364	0.88153
	Equal variances not assumed#			0.984	8.997	0.351	0.32895	0.33430	−0.42733	1.08523
IHRM3	Equal variances assumed#	3.763	0.064	−1.025	25	0.315	−0.27632	0.26970	−0.83176	0.27913
	Equal variances not assumed			−1.208	19.738	0.241	−0.27632	0.22876	−0.75391	0.20128
IHRM4	Equal variances assumed	13.683	0.001	1.585	25	0.125	0.12500	0.07885	−0.03739	0.28739
	Equal variances not assumed#			1.000	7.000	0.351	0.12500	0.12500	−0.17058	0.42058
*IHRM5	Equal variances assumed#	0.342	0.564	4.487	25	0.000	1.69079	0.37678	0.91479	2.46679
	Equal variances not assumed			4.958	16.752	0.000	1.69079	0.34101	0.97051	2.41107
IHRM6	Equal variances assumed	6.107	0.021	−1.014	25	0.320	−0.19079	0.18811	−0.57821	0.19663
	Equal variances not assumed#			−1.148	17.801	0.266	−0.19079	0.16622	−0.54028	0.15870
IHRM7	Equal variances assumed#	2.424	0.132	1.536	25	0.137	0.40132	0.26133	−0.13691	0.93954
	Equal variances not assumed			1.263	9.313	0.237	0.40132	0.31765	−0.31360	1.11623
ISCM1	Equal variances assumed#	1.357	0.255	−0.763	25	0.453	−0.31579	0.41385	−1.16814	0.53656
	Equal variances not assumed			−0.725	11.874	0.483	−0.31579	0.43583	−1.26650	0.63492
ISCM2	Equal variances assumed	4.587	0.042	−0.157	25	0.876	−0.07237	0.46032	−1.02042	0.87568
	Equal variances not assumed#			−0.229	23.224	0.821	−0.07237	0.31665	−0.72706	0.58232
*ISCM3	Equal variances assumed	8.356	0.008	−1.334	25	0.194	−0.34868	0.26133	−0.88691	0.18954
	Equal variances not assumed#			−1.718	23.829	0.099	−0.34868	0.20291	−0.76762	0.07025
ISCM4	Equal variances assumed#	2.298	0.142	−0.583	25	0.565	−0.17105	0.29360	−0.77572	0.43362
	Equal variances not assumed			−0.711	21.444	0.485	−0.17105	0.24057	−0.67071	0.32860

(Continued)

Table 6.14 (Continued)

		Levene's test for equality of variances		t test for equality of means						
		F	Sig.	*t*	df	Sig. (two-tailed)	Mean difference	Std. error difference	95 % confidence interval of the difference	
									Lower	Upper
ISCM5	Equal variances assumed#	2.101	0.160	0.094	25	0.926	0.0395	0.4186	−0.8226	0.9015
	Equal variances not assumed			0.068	7.955	0.947	0.0395	0.5774	−1.2934	1.3723
ISCM6	Equal variances assumed#	1.688	0.206	−0.577	25	0.569	−0.11842	0.20511	−0.54086	0.30401
	Equal variances not assumed			−0.594	14.109	0.562	−0.11842	0.19928	−0.54553	0.30868
ISCM7	Equal variances assumed#	0.035	0.852	−0.616	25	0.543	−0.19737	0.32039	−0.85723	0.46250
	Equal variances not assumed			−0.642	14.525	0.531	−0.19737	0.30742	−0.85450	0.45976
ISCM8	Equal variances assumed#	0.736	0.399	−1.452	25	0.159	−0.77632	0.53469	−1.87753	0.32490
	Equal variances not assumed			−1.314	10.848	0.216	−0.77632	0.59086	−2.07901	0.52638
ISCM9	Equal variances assumed	8.356	0.008	−1.334	25	0.194	−0.34868	0.26133	−0.88691	0.18954
	Equal variances not assumed#			−1.718	23.829	0.099	−0.34868	0.20291	−0.76762	0.07025
ISM1	Equal variances assumed#	0.585	0.452	−0.544	25	0.591	−0.21711	0.39896	−1.03877	0.60456
	Equal variances not assumed			−0.618	17.968	0.544	−0.21711	0.35124	−0.95512	0.52091
ISM2	Equal variances assumed#	0.089	0.768	−0.838	25	0.410	−0.32237	0.38464	−1.11455	0.46981
	Equal variances not assumed			−0.935	17.188	0.362	−0.32237	0.34460	−1.04881	0.40407
ISM3	Equal variances assumed#	0.253	0.619	−0.440	25	0.663	−0.15132	0.34356	−0.85888	0.55625
	Equal variances not assumed			−0.464	14.886	0.649	−0.15132	0.32625	−0.84716	0.54453
ISM4	Equal variances assumed#	0.665	0.422	−1.702	25	0.101	−0.60526	0.35553	−1.33748	0.12696
	Equal variances not assumed			−2.115	22.302	0.046	−0.60526	0.28615	−1.19824	−0.01229
IMM1	Equal variances assumed#	3.523	0.072	0.664	25	0.513	0.19737	0.29739	−0.41512	0.80986
	Equal variances not assumed			0.573	10.028	0.579	0.19737	0.34428	−0.56944	0.96418
IMM2	Equal variances assumed#	1.035	0.319	0.107	25	0.916	0.03289	0.30858	−0.60263	0.66842
	Equal variances not assumed			0.115	15.713	0.910	0.03289	0.28656	−0.57549	0.64128

(Continued)

Table 6.14 (Continued)

		Levene's test for equality of variances		t test for equality of means						
		F	Sig.	*t*	df	Sig. (two-tailed)	Mean difference	Std. error difference	95 % confidence interval of the difference	
									Lower	Upper
IMM3	Equal variances assumed#	0.227	0.638	−0.061	25	0.952	−0.01974	0.32279	−0.68454	0.64507
	Equal variances not assumed			−0.058	11.861	0.955	−0.01974	0.34011	−0.76174	0.72227
IMM4	Equal variances assumed#	0.113	0.739	−0.292	25	0.773	−0.09211	0.31569	−0.74228	0.55807
	Equal variances not assumed			−0.302	14.248	0.767	−0.09211	0.30542	−0.74609	0.56188
IMM5	Equal variances assumed#	3.039	0.094	−0.829	25	0.415	−0.22368	0.26970	−0.77913	0.33176
	Equal variances not assumed			−0.978	19.738	0.340	−0.22368	0.22876	−0.70128	0.25391
IMM6	Equal variances assumed#	4.052	0.055	0.801	25	0.431	0.38158	0.47624	−0.59926	1.36242
	Equal variances not assumed			0.670	9.528	0.519	0.38158	0.56991	−0.89685	1.66001
ICM1	Equal variances assumed#	0.800	0.380	−0.176	25	0.862	−0.05921	0.33696	−0.75320	0.63478
	Equal variances not assumed			−0.161	11.154	0.875	−0.05921	0.36669	−0.86492	0.74650
ICM2	Equal variances assumed#	0.002	0.961	0.253	25	0.802	0.08553	0.33752	−0.60960	0.78066
	Equal variances not assumed			0.247	12.545	0.809	0.08553	0.34595	−0.66461	0.83566
ICM3	Equal variances assumed#	0.243	0.626	0.189	25	0.852	0.05263	0.27890	−0.52177	0.62703
	Equal variances not assumed			0.174	11.196	0.865	0.05263	0.30289	−0.61261	0.71787
ICM4	Equal variances assumed#	1.440	0.241	0.491	25	0.627	0.15789	0.32127	−0.50377	0.81956
	Equal variances not assumed			0.434	10.417	0.673	0.15789	0.36342	−0.64748	0.96327
ICM5	Equal variances assumed#	1.206	0.283	−0.894	25	0.380	−0.38816	0.43441	−1.28284	0.50652
	Equal variances not assumed			−1.001	17.334	0.331	−0.38816	0.38789	−1.20533	0.42902
ICM6	Equal variances assumed#	2.371	0.136	−0.760	25	0.454	−0.34211	0.45021	−1.26933	0.58511
	Equal variances not assumed			−1.016	24.847	0.319	−0.34211	0.33657	−1.03549	0.35128
ICM7	Equal variances assumed#	1.646	0.211	−0.746	25	0.462	−0.20395	0.27323	−0.76667	0.35877
	Equal variances not assumed			−0.842	17.634	0.411	−0.20395	0.24232	−0.71381	0.30591

Note The letter "I" in front of each of the factors refers to "Importance" of that respective factor; hash symbols indicate the values which were used; asterisks indicate the factors which have a significantly different mean rating as determined by the A1 and A2 contractors

6.7.2 Usage Rating

In contrast, Table 6.15 shows that for all 33 CSFs, the mean degree of usage by the A1 contractors is higher than the A2 contractors. The relatively lower levels of CONQUAS management practices by A2 contractors as compared to A1 contractors are expected because based on their average CONQUAS scores obtained, A1 contractors have a higher CONQUAS score than A2 contractors in general. This result provides some indication that there is a difference in the level of CONQUAS management practices among contractors of different financial grade. This also reflects that A1 contractors have a higher awareness of adopting the 33 CSFs than A2 contractors. This lack of awareness may be because A2 contractors, being financially constrained (Campello et al. 2010), are more prone to be interested in completing the project on time and within budget. They do not appear to care about the working system, documentation and proper CONQUAS management procedures required to attain a high CONQUAS score. They may think that it is pointless to change their quality practices for tasks which they have been doing every day for many years (Low 1998).

However, even within the same firm, this level of awareness may vary because when individuals from different perspectives are put together to form the QMU, differences in opinions of which CSFs are important and which CSFs should not be implemented can occur (Collins 2005). Furthermore, each person may prefer to stick to their routine method of handling workmanship quality issues. This may result in inefficient workflow as well as poor communication within the QMU to come together and solve problems. Hence, adequate preparation is required to assure that the QMU will be able to manage the CONQUAS workflow effectively as a team and periodic reinforcement is also essential in order to cement the CONQUAS management framework throughout the entire organisation. As clarified by Ms B:

> I won't say that CONQUAS management is consistent throughout all our projects as different projects are made up of different teams of people who may have their own way of doing things.

Next, out of 33 CSFs, the mean usage rating of thirteen CSFs (indicated with an asterisk) of the two groups is statistically significantly different as shown in Table 6.16. Again, it is observed from both Tables 6.15 and 6.16 that there is no noticeable trend to elaborate on the results obtained. It can only be inferred that A2 contractors have to put in much more effort and resources to implement these thirteen statistically significant CSFs which were highly utilised by the A1 contractors in order to improve their CONQUAS scores. This can be explained by the fact that the management and development requirements of A1 contractors require twelve more professionals with qualifications recognised by the Professional Engineers Board of Singapore, Board of Architects of Singapore, and BCA-recognised resident engineer or professional or technical personnel with relevant qualifications, as compared to the A2 contractors (BCA 2012). Hence, A1 contractors seem to have a better team of staffs to deal with the CONQUAS management activities, while A2 contractors do not usually employ such building professionals to ensure good workmanship quality.

Table 6.15 Group statistics for degree of usage

		N	Mean	Std. deviation	Std. error mean
UHRM1	A1	19	1.0000	0.00000	0.00000
	A2	8	0.6250	0.51755	0.18298
UHRM2	A1	19	0.7895	0.41885	0.09609
	A2	8	0.3750	0.51755	0.18298
UHRM3	A1	19	0.8421	0.37463	0.08595
	A2	8	0.3750	0.51755	0.18298
UHRM4	A1	19	1.0000	0.00000	0.00000
	A2	8	0.6250	0.51755	0.18298
UHRM5	A1	19	0.2105	0.41885	0.09609
	A2	8	0.0000	0.00000	0.00000
UHRM6	A1	19	0.7895	0.41885	0.09609
	A2	8	0.3750	0.51755	0.18298
UHRM7	A1	19	1.0000	0.00000	0.00000
	A2	8	0.6250	0.51755	0.18298
USCM1	A1	19	0.8421	0.37463	0.08595
	A2	8	0.5000	0.53452	0.18898
USCM2	A1	19	0.8421	0.37463	0.08595
	A2	8	0.3750	0.51755	0.18298
USCM3	A1	19	0.8947	0.31530	0.07234
	A2	8	0.5000	0.53452	0.18898
USCM4	A1	19	1.0000	0.00000	0.00000
	A2	8	0.6250	0.51755	0.18298
USCM5	A1	19	0.7895	0.41885	0.09609
	A2	8	0.6250	0.51755	0.18298
USCM6	A1	19	0.8421	0.37463	0.08595
	A2	8	0.3750	0.51755	0.18298
USCM7	A1	19	1.0000	0.00000	0.00000
	A2	8	0.6250	0.51755	0.18298
USCM8	A1	19	0.3158	0.47757	0.10956
	A2	8	0.2500	0.46291	0.16366
USCM9	A1	19	0.8947	0.31530	0.07234
	A2	8	0.6250	0.51755	0.18298
USM1	A1	19	1.0000	0.00000	0.00000
	A2	8	0.6250	0.51755	0.18298
USM2	A1	19	1.0000	0.00000	0.00000
	A2	8	0.6250	0.51755	0.18298
USM3	A1	19	0.5789	0.50726	0.11637
	A2	8	0.1250	0.35355	0.12500
USM4	A1	19	0.8947	0.31530	0.07234
	A2	8	0.3750	0.51755	0.18298
UMM1	A1	19	0.3158	0.47757	0.10956
	A2	8	0.2500	0.46291	0.16366
UMM2	A1	19	1.0000	0.00000	0.00000
	A2	8	0.6250	0.51755	0.18298
UMM3	A1	19	0.8947	0.31530	0.07234
	A2	8	0.6250	0.51755	0.18298

(Continued)

Table 6.15 (Continued)

		N	Mean	Std. deviation	Std. error mean
UMM4	A1	19	0.6842	0.47757	0.10956
	A2	8	0.2500	0.46291	0.16366
UMM5	A1	19	0.7895	0.41885	0.09609
	A2	8	0.3750	0.51755	0.18298
UMM6	A1	19	0.3684	0.49559	0.11370
	A2	8	0.1250	0.35355	0.12500
UCM1	A1	19	0.8947	0.31530	0.07234
	A2	8	0.6250	0.51755	0.18298
UCM2	A1	19	0.6316	0.49559	0.11370
	A2	8	0.3750	0.51755	0.18298
UCM3	A1	19	0.4737	0.51299	0.11769
	A2	8	0.2500	0.46291	0.16366
UCM4	A1	19	0.8947	0.31530	0.07234
	A2	8	0.3750	0.51755	0.18298
UCM5	A1	19	0.7368	0.45241	0.10379
	A2	8	0.1250	0.35355	0.12500
UCM6	A1	19	0.2105	0.41885	0.09609
	A2	8	0.0000	0.00000	0.00000
UCM7	A1	19	0.9474	0.22942	0.05263
	A2	8	0.6250	0.51755	0.18298

Note The letter "U" in front of each of the factors refers to "Usage" of that respective factor

As a whole, since the means of twenty CSFs of the two groups are not significantly different from each other, the null hypothesis $H_{7.0}$ would not be rejected, indicating that in general, the degree of usage by A1 and A2 contractors is about the same. Therefore, this accepts the null hypothesis $H_{7.0}$ and affirms the notion that the different responses received by A1 and A2 contractors do not affect the usage level of the CSFs received to a large extent.

6.8 Summary of Analysis

Figure 6.8 shows the matrix diagram of the 33 CSFs identified. For the majority of the CSFs, although high importance means high usage, the success of such adoption is still unknown as some may feel that they have been successful in practicing these CSF, but a third party may feel that they have not. This may be the reason for contractors not being able to achieve good CONQUAS scores as they are immersed in their current state of executing the CONQUAS management workflow and have not reflected whether they are doing it the correct way and successfully. To add on, it is upsetting that when contractors perceived that the CSFs are of high importance, yet they do not adopt them. This may be due to the lack of resources and expertise to do so but should not be an excuse for not

Table 6.16 Independent sample test for degree of usage

		Levene's test for equality of variances		*t* test for equality of means						
		F	Sig.	*t*	df	Sig. (two-tailed)	Mean difference	Std. error difference	95 % confidence interval of the difference	
									Lower	Upper
UHRM1	Equal variances assumed	263.889	0.000	3.249	25	0.003	0.37500	0.11542	0.13728	0.61272
	Equal variances not assumed#			2.049	7.000	0.080	0.37500	0.18298	−0.05768	0.80768
*UHRM2	Equal variances assumed#	2.225	0.148	2.192	25	0.038	0.41447	0.18910	0.02501	0.80394
	Equal variances not assumed			2.005	11.066	0.070	0.41447	0.20668	−0.04009	0.86904
*UHRM3	Equal variances assumed	4.454	0.045	2.641	25	0.014	0.46711	0.17684	0.10290	0.83131
	Equal variances not assumed#			2.311	10.236	0.043	0.46711	0.20216	0.01806	0.91615
UHRM4	Equal variances assumed	263.889	0.000	3.249	25	0.003	0.37500	0.11542	0.13728	0.61272
	Equal variances not assumed#			2.049	7.000	0.080	0.37500	0.18298	−0.05768	0.80768
*UHRM5	Equal variances assumed	14.692	0.001	1.405	25	0.172	0.21053	0.14979	−0.09798	0.51903
	Equal variances not assumed#			2.191	18.000	0.042	0.21053	0.09609	0.00865	0.41241
*UHRM6	Equal variances assumed#	2.225	0.148	2.192	25	0.038	0.41447	0.18910	0.02501	0.80394
	Equal variances not assumed			2.005	11.066	0.070	0.41447	0.20668	−0.04009	0.86904
UHRM7	Equal variances assumed	263.889	0.000	3.249	25	0.003	0.37500	0.11542	0.13728	0.61272
	Equal variances not assumed#			2.049	7.000	0.080	0.37500	0.18298	−0.05768	0.80768
USCM1	Equal variances assumed	6.520	0.017	1.908	25	0.068	0.34211	0.17933	−0.02724	0.71145
	Equal variances not assumed#			1.648	10.028	0.130	0.34211	0.20761	−0.12030	0.80451
*USCM2	Equal variances assumed	4.454	0.045	2.641	25	0.014	0.46711	0.17684	0.10290	0.83131
	Equal variances not assumed#			2.311	10.236	0.043	0.46711	0.20216	0.01806	0.91615
USCM3	Equal variances assumed	12.255	0.002	2.406	25	0.024	0.39474	0.16409	0.05679	0.73268
	Equal variances not assumed#			1.951	9.125	0.082	0.39474	0.20235	−0.06206	0.85154
USCM4	Equal variances assumed	263.889	0.000	3.249	25	0.003	0.37500	0.11542	0.13728	0.61272
	Equal variances not assumed#			2.049	7.000	0.080	0.37500	0.18298	−0.05768	0.80768

(Continued)

Table 6.16 (Continued)

		Levene's test for equality of variances		t test for equality of means						
		F	Sig.	*t*	df	Sig. (two-tailed)	Mean difference	Std. error difference	95 % confidence interval of the difference	
									Lower	Upper
USCM5	Equal variances assumed#	2.225	0.148	0.870	25	0.393	0.16447	0.18910	−0.22499	0.55394
	Equal variances not assumed			0.796	11.066	0.443	0.16447	0.20668	−0.29009	0.61904
*USCM6	Equal variances assumed	4.454	0.045	2.641	25	0.014	0.46711	0.17684	0.10290	0.83131
	Equal variances not assumed#			2.311	10.236	0.043	0.46711	0.20216	0.01806	0.91615
USCM7	Equal variances assumed	263.889	0.000	3.249	25	0.003	0.37500	0.11542	0.13728	0.61272
	Equal variances not assumed#			2.049	7.000	0.080	0.37500	0.18298	−0.05768	0.80768
USCM8	Equal variances assumed#	0.493	0.489	0.330	25	0.744	0.06579	0.19957	−0.34523	0.47680
	Equal variances not assumed			0.334	13.616	0.743	0.06579	0.19695	−0.35775	0.48933
USCM9	Equal variances assumed	8.977	0.006	1.672	25	0.107	0.26974	0.16136	−0.06259	0.60206
	Equal variances not assumed#			1.371	9.271	0.203	0.26974	0.19676	−0.17339	0.71287
USM1	Equal variances assumed	263.889	0.000	3.249	25	0.003	0.37500	0.11542	0.13728	0.61272
	Equal variances not assumed#			2.049	7.000	0.080	0.37500	0.18298	−0.05768	0.80768
USM2	Equal variances assumed	263.889	0.000	3.249	25	0.003	0.37500	0.11542	0.13728	0.61272
	Equal variances not assumed#			2.049	7.000	0.080	0.37500	0.18298	−0.05768	0.80768
*USM3	Equal variances assumed	16.733	0.000	2.295	25	0.030	0.45395	0.19780	0.04657	0.86133
	Equal variances not assumed#			2.658	18.878	0.016	0.45395	0.17079	0.09633	0.81156
*USM4	Equal variances assumed	8.977	0.006	3.221	25	0.004	0.51974	0.16136	0.18741	0.85206
	Equal variances not assumed#			2.641	9.271	0.026	0.51974	0.19676	0.07661	0.96287
UMM1	Equal variances assumed#	0.493	0.489	0.330	25	0.744	0.06579	0.19957	−0.34523	0.47680
	Equal variances not assumed			0.334	13.616	0.743	0.06579	0.19695	−0.35775	0.48933
UMM2	Equal variances assumed	263.889	0.000	3.249	25	0.003	0.37500	0.11542	0.13728	0.61272
	Equal variances not assumed#			2.049	7.000	0.080	0.37500	0.18298	−0.05768	0.80768

(Continued)

Table 6.16 (Continued)

		Levene's test for equality of variances		*t* test for equality of means						
		F	Sig.	*t*	df	Sig. (two-tailed)	Mean difference	Std. error difference	95 % confidence interval of the difference	
									Lower	Upper
UMM3	Equal variances assumed	8.977	0.006	1.672	25	0.107	0.26974	0.16136	−0.06259	0.60206
	Equal variances not assumed#			1.371	9.271	0.203	0.26974	0.19676	−0.17339	0.71287
*UMM4	Equal variances assumed#	0.493	0.489	2.176	25	0.039	0.43421	0.19957	0.02320	0.84523
	Equal variances not assumed			2.205	13.616	0.045	0.43421	0.19695	0.01067	0.85775
*UMM5	Equal variances assumed#	2.225	0.148	2.192	25	0.038	0.41447	0.18910	0.02501	0.80394
	Equal variances not assumed			2.005	11.066	0.070	0.41447	0.20668	−0.04009	0.86904
UMM6	Equal variances assumed	10.722	0.003	1.255	25	0.221	0.24342	0.19398	−0.15610	0.64294
	Equal variances not assumed#			1.441	18.460	0.166	0.24342	0.16897	−0.11095	0.59779
UCM1	Equal variances assumed	8.977	0.006	1.672	25	0.107	0.26974	0.16136	−0.06259	0.60206
	Equal variances not assumed#			1.371	9.271	0.203	0.26974	0.19676	−0.17339	0.71287
UCM2	Equal variances assumed#	0.004	0.951	1.213	25	0.236	0.25658	0.21151	−0.17903	0.69219
	Equal variances not assumed			1.191	12.712	0.255	0.25658	0.21543	−0.20990	0.72306
UCM3	Equal variances assumed	5.541	0.027	1.063	25	0.298	0.22368	0.21051	−0.20987	0.65724
	Equal variances not assumed#			1.110	14.593	0.285	0.22368	0.20158	−0.20703	0.65440
*UCM4	Equal variances assumed	8.977	0.006	3.221	25	0.004	0.51974	0.16136	0.18741	0.85206
	Equal variances not assumed#			2.641	9.271	0.026	0.51974	0.19676	0.07661	0.96287
*UCM5	Equal variances assumed#	3.050	0.093	3.399	25	0.002	0.61184	0.17998	0.24116	0.98253
	Equal variances not assumed			3.766	16.862	0.002	0.61184	0.16247	0.26884	0.95484
*UCM6	Equal variances assumed	14.692	0.001	1.405	25	0.172	0.21053	0.14979	−0.09798	0.51903
	Equal variances not assumed#			2.191	18.000	0.042	0.21053	0.09609	0.00865	0.41241
UCM7	Equal variances assumed	21.889	0.000	2.276	25	0.032	0.32237	0.14161	0.03072	0.61402
	Equal variances not assumed#			1.693	8.184	0.128	0.32237	0.19040	−0.11498	0.75972

Note The letter "U" in front of each of the factors refers to "Usage" of that respective factor; hash symbols indicate the values which were used; asterisks indicate the factors which have a significantly different mean rating as determined by the A1 and A2 contractors

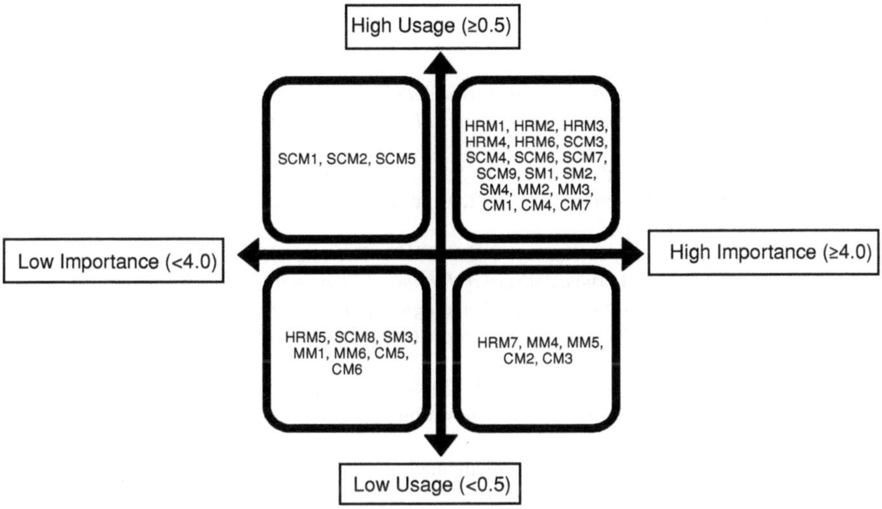

Fig. 6.8 Matrix diagram of the 33 CSFs

performing as the contractors should have mulled over such problems and find the solutions to implementing these important CSFs in order to reap the benefits of the CONQUAS management framework.

Next, although the predictability of the regression equation is high, being adequate to predict CONQUAS score on its own as further supported by the interviewees, this regression model is only at its exploratory stage and hence suggesting that it is still not of sufficient rigour yet. This is because the sample size is relatively small and there may also be insufficient variables tested as pointed out by the interviewees. Moreover, the experience of the personnel involved in the CONQUAS workflow, discipline of personnel, frequency or percentage of CONQUAS projects the firm was involved with were not taken into account to verify whether they might also have influenced the CONQUAS scores. Furthermore, the duration of the contract, contract sum involved, gross floor area of the project as well as the development type might have also affected the results obtained.

To reiterate, the hypotheses tested have given this research a boost to develop a coherent CONQUAS management framework which interlinks the various CSFs together. Also, regardless of the financial grade category the contractors belong to, the proposed framework constructed for contractors to consider to achieving high CONQUAS scores will not be affected as determined by the independent *t* tests. Therefore, after analysing the quantitative data and understanding the trends, the results suggest that it is imperative to further conduct an in-depth case study to understand the benefits in specific (qualitative) details. A pre–post case study will be used to draw a more substantive conclusion in the next chapter.

References

Anink, D., Boonstra, C., & Mak, J. (1996). *Handbook of sustainable building: an environmental preference method for selection of materials for use in construction and refurbishment.* London: Earthscan.

Building and Construction Authority. (2012). *Specific registration requirements for construction workhead.* Retrieved November 18, 2012, from http://www.bca.gov.sg/ContractorsRegistry/others/Registration_CW.pdf

Campello, M., Graham, J. R., & Harvey, C. R. (2010). The real effects of financial constraints: Evidence from a financial crisis. *Journal of Financial Economics, 97*(3), 470–487.

Collins, D. (2005). *Organisational change: Sociological perspectives*. London: Routledge.

Culp, G. L., & Smith, A. (1992). *Managing People (Including Yourself) for Project Success*. New York: Wiley.

Eiadat, Y., Kelly, A., Roche, F., & Eyadat, H. (2008). Green and competitive? An empirical test of the mediating role of environmental innovation strategy. *Journal of World Business, 43*(2), 131–145.

Goh, P. L., & Ridgway, K. (1994). The implementation of total quality management in small and medium-sized manufacturing companies. *The TQM magazine, 6*(2), 54–60.

Haksever, C. (1996). Total quality management in the small business environment. *Business Horizons, 39*(2), 33–40.

Hamel, G., & Prahalad, C. K. (1994). Competing for the Future. *Harvard Business Review, 72*(4), 122–128.

Hendricks, C. F. (1992). *The rightsizing remedy: How managers can respond to the downsizing dilemma*. Illinois: Irwin Professional Pub.

Hubbard, C., Scow, R., & Texas, S. (1990). Succession in the small business the firm's future and management's fears. *Proceedings, SBIDA* (pp. 147–151).

Koh, T. Y., & Low, S. P. (2008). Organizational culture and TQM implementation in construction firms in Singapore. *Construction Management and Economics, 26*(3), 237–248.

Laufer, A., & Tucker, R. L. (1987). Is construction project planning really doing its job? A critical examination of focus, role and process. *Construction Management and Economics, 5*(3), 243–266.

Leonard, F. S., & Sasser, W. E. (1982). The incline of quality. *Harvard Business Review, 60*(5), 163–171.

Low, S. P. (1998). *ISO 9000 and the construction industry*. Chartridge Books, Oxford.

Low, S. P., & Teo, J. (2004). Implementing total quality management in construction firms. *Journal of Management in Engineering, 20*(1), 8–15.

Maloney, W. F. (1997). Strategic planning for human resource management in construction. *Journal of Management in Engineering, 13*(3), 49–56.

Poon, S. W., & Xu, Y. Q. (1997). Problems of small Chinese contractors in Hong Kong during quality assurance system implementation. In *Proceedings of International Conference on Leadership and Total Quality Management in Construction and Building* (pp. 6–8).

Rao, A., Thornberry, N., & Weintraub, J. (1987). An empirical study of autonomous work groups relationships between worker reactions and effectiveness. *Behavioral Science, 32*(1), 66–76.

Reed, D. H., & Frankham, R. (2003). Correlation between fitness and genetic diversity. *Conservation Biology, 17*(1), 230–237.

Saraph, J. V., Benson, P. G., & Schroeder, R. G. (1989). An instrument for measuring the critical factors of quality management. *Decision Sciences, 20*(4), 810–829.

Shea, J., & Gobeli, D. (1995). TQM: The experiences of ten small businesses. *Business Horizons, 38*(1), 71–77.

Wang, R. Y. (1998). A product perspective on total data quality management. *Communications of the ACM, 41*(2), 58–65.

Chapter 7
Case Study

7.1 Outline of Company Z

Company Z was founded in 1983 and is now a reputable grade A1 (general building) main contractor headquartered in Singapore. Company Z is dedicated to provide reliable, timely, excellent and safe construction services to their clients. Company Z also envisions to being the contractor of choice and aims to achieve total customer satisfaction through quality construction and services, timely completion of projects as well as provide a safe and healthy working environment. Moreover, Company Z strives to work with passion, commitment and enthusiasm; honesty and integrity; continual self-improvement in skill and knowledge; and excellence in reputation. Be it cash grants, donations, sponsorships or in-kind services, Company Z has been supporting those in need and also other worthy causes. They also participate in voluntary work both locally and overseas by lending their knowledge and expertise as they believe in bringing corporate social responsibility to a personal level. Company Z has consistently exceeded industry norms with the aim of reducing the harmful effects from its operations. They are committed to reducing them by evaluating their operations and systems to ensure that they are efficient and able to achieve both quality and productivity in their work processes.

Having almost 30 years of experience, Company Z has successfully completed an extensive range of projects worth well over S$2.5 billion including the following: condominiums, landed properties, office buildings, shopping malls, highways, hospitals, clubhouses, international resorts, embassies, hotels, churches, schools, road works, bridges, highways, mass rapid transit or viaduct, power station structures, sewerage and water treatment plants, workshops and industrial buildings, as well as heavy civil engineering works. Company Z also provides a whole range of construction services including the following: piling and foundation, structural works, architecture works, M&E works, interior design works as well as hardscaping and softscaping. Company Z's core competencies are in project management and control, coordination, detailing and design, supervision and quality control, construction site, as well

S. P. Low and J. Ong, *Project Quality Management*, DOI: 10.1007/978-981-287-074-2_7,

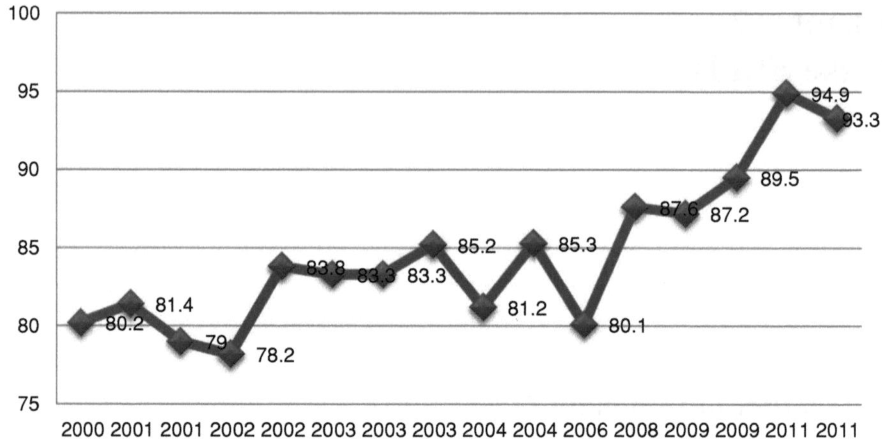

Fig. 7.1 CONQUAS scores over the years. *Source* Corenet (2012)

as risk and safety management. A strong financial standing, wide network of connections and involvement in all kinds and types of projects are their core strengths.

Over the years, Company Z has garnered multiple awards in both local and international competitions. They feel honoured to have their achievements and efforts recognised in the form of numerous industry awards as well as by their peers. Specifically, the BCA CEA is viewed as an important benchmark for Company Z to work towards to enhance their work performance, particularly in the workmanship quality management aspect (CONQUAS), which accounts for almost 50 % in the CEA evaluation criteria. Moreover, participation in the BCA QM for Good Workmanship Scheme will be taken into consideration for all private residential and mixed development projects. Additional points will be awarded for projects that have successfully completed the QM assessment, as it is a sign of demonstrating a higher consistency in quality achievement.

7.2 CONQUAS Background

The CONQUAS score profile of all their sixteen CONQUAS 21 projects is charted in Fig. 7.1. It is observed that Company Z's CONQUAS score was not very stable in the beginning (year 2000) until the spike in year 2008, from 80.1 to 87.6, demonstrating a noteworthy improvement in their construction workmanship quality standard. Following this assertion, it is reckoned that the substantially different score profile before and after year 2008 is due to the different usage level of the 33 CSFs as underpinned in Chap. 4. Therefore, the following sections will seek to gain a better understanding of the different CONQUAS management practices in these two phases and to back up the results of the analysis made in Chap. 6.

7.3 Early Period (Before Year 2002)

In the beginning, Company Z's CONQUAS management procedures and processes were relatively simple. There was no official quality management team set up for each project, and there was also a great deal of uncertainty around how and who would execute the CONQUAS management practices to assure good construction workmanship standard. As a result, this uncertainty led to inaction as staffs were left with a vague sense of having to do something but with no clear guidelines on how they might carry out the work. Furthermore, subcontractors and suppliers were selected based on the lowest cost approach and only minimal workmanship quality standard was needed. Moreover, the format of the subcontract was simple, and sometimes, only verbal contracts were relied upon for certain conditions. When defects were found, corrective actions were taken, but nothing was done to prevent such instances from occurring again, and the same practices were still used. Essentially, this did not encourage or result in any quality improvement and effective quality policies, and goals were absent as well. Even if Company Z repeatedly incurs additional costs due to defects for particular trade types, they did not use this knowledge and the experience gained to include a contingency amount in future contracts for potential defects.

7.4 Transition Period (Year 2002–2008)

Continuing with such a practice in the short term may be profitable, but in the long term, Company Z realised that this action has had detrimental effects on their ability to deliver quality projects to their clients' satisfaction. Thereafter, in 2002, Company Z started to modify their CONQUAS management framework. Gradually, Company Z grew to become a trusted partner of many of the biggest property developers in Singapore. Company Z started to participate actively in various construction projects in and around the region. This was coupled with Company Z's vast and diverse experience since their inception, which gave them the necessary prerequisites to undertake bigger and more complicated projects both locally and overseas.

But these undertakings only started to achieve recognition after years of planning and numerous adjustments to their CONQUAS management framework, as well as the hard work put in. This may be due to the fact that the final score would only be known at least one year later after project completion depending on when CONQUAS assessment takes place or even up to three years depending on the size of the project and hence slowing down the learning curve process. Out of 16 CONQUAS 21 projects that they have undertaken, nine of them have won the CEA (three excellent and six merit), meaning that these projects have attained the minimum CONQUAS scores for the respective category listed that year. This is an outstanding achievement which implies that these projects have been accorded recognition by the principal consultant to the project owner, demonstrating the highest standards of quality performance excellence in the Singapore construction industry. The three "Excellent" accomplishments did not come easy and only occurred in years 2008, 2009 and 2011 which suggests that much time and effort have been put in to attain excellent workmanship quality, to achieve a high CONQUAS score. Effectively, the CEA was seen as a critical push factor to spur them to improve their work performance and compete at the national level.

Basically, this transition started when the top management of Company Z decided to change their existing mindset and incorporate a comprehensive CONQUAS management framework to become a part of their organisational life. It was only then can good workmanship quality be implemented successfully to attain high CONQUAS scores. Company Z has confirmed the usage of all the 33 CSFs pointed out in Chap. 4 as part of their renewed CONQUAS management framework, indicating that they should be able to achieve the maximum possible score (93.4) based on the proposed linear regression equation. This score was found to be approximately accurate to the CONQUAS score obtained in their last two CONQUAS projects completed in 2011. Moreover, they are confident of achieving at least 90 points in their current and future projects and have input this target score into their corporate quality policy.

7.5 Current Practices (Year 2008 Onwards)

The following will elaborate on those five major strategies which are of great concern to Company Z in meeting their CONQUAS score target. These strategies also replicate some of the CSFs identified in this study.

(1) Company Z employs various systematic problem-solving techniques. Benchmarking is used to determine the quality performance of the built product. Failure mode effect analysis is used to identify the causes of defects and to take corrective actions immediately as well as experimenting with new approaches to prevent such defects from occurring again. In particular, pre-construction activities are mapped out with all activities being listed and charted in diagrams to assist in the compilation of the data and to help stimulate further discussion. The data are finally sorted into priority items so that the most important ones received the attention that they required. Thereafter, a detailed process chart is used, showing a breakdown of each activity, its relationship with other activities and the constraints imposed upon it and, in addressing whether these constraints can be removed or how they can be made easier, enabling solutions to the workmanship quality issues to be found. Finally, once a new method of working has been agreed, it is tested and monitored to see its effectiveness in improving the CONQUAS performance.

(2) Company Z also learns from their own experience and past history by implementing a methodology for generating feedback to engender learning in projects. This is mainly used to identify problematic trades where, for example, there has been a high rate of defect non-compliance. Similarly, Company Z also gains experiences through inter-organisational partnering with other contractors so as to promote learning at the individual, team and organisational level to enhance their quality performance. In addition, Company Z learns

from the experiences and best practices of others. They have realised that TQM has matured in the manufacturing industry, and thus, besides taking into account the practices of the construction industry, they believe in applying the experiences of these organisations into their CONQUAS management framework to overcome the difficulties faced.

(3) Furthermore, Company Z has a list of training programmes to transfer quality management knowledge quickly and efficiently throughout the organisation. Staffs are encouraged to engage in training programs to raise awareness of good industry best practices to achieve superior workmanship quality. There are regular reviews with the QMU which are conducted to identify workmanship quality failures and successes of the projects and enable employees to hear about each other's experiences on particular projects.

(4) Company Z has also designed checklists for each trade of works to be used at different stages of inspections, and five of these inspection checklist templates are shown in Appendix G. The use of such checklists enables them to pay attention to all details required to achieve good workmanship quality rather than simply relying on their experiences and memory to spot any signs of defect non-compliance. Most importantly, the checklists used are unique to each project to make sure that they are able to meet the quality requirements as stipulated in the respective contracts.

(5) Lastly, on top of past coordination and relationship, Company Z only considers subcontractors and suppliers who are ISO 9001 certified as they feel that the journey towards achieving high CONQUAS scores will require not only the full commitment of the main contractor, but also having a basic quality culture created externally by the subcontractors and suppliers who genuinely understand the QMS and share similar quality values in their operations and product offerings.

In fact, Company Z also attributes their success to the Singapore government who has been very proactive in promoting quality. Their role in supporting the quality movement has had a positive impact in helping to inculcate quality consciousness throughout the entire company and workers. They believe that government initiatives have played, and will continue to play, a key role in helping the local construction industry progress towards quality class companies. Together with the design and implementation of several prestigious national and regional quality awards such as the Deming Prize, European Quality Award and Singapore Quality Award, this has spurred Company Z to do their best to achieve workmanship quality excellence.

7.6 Outcome

Using these guidelines as well as the 33 CSFs, Company Z was able to acquire a reputation for delivering high-quality building and services as well as enhancing productivity and achieving more sustainable growth. Such an improvement to their workmanship quality has also resulted in improved productivity and more satisfied clients with greater loyalty, increased sales and an enhanced competitive position. Company Z believes in satisfying their clients to excel in today's exceedingly competitive markets. Moreover, these practices have reduced the rework they experienced in projects from about 5 % to less than 1 % of the contract value through effectively implementing a QA system in conjunction with continuous quality improvement practices.

After implementing this new CONQUAS management system, Company Z has improved its workmanship quality standard and at the same time reduced the cost of poor quality, leading to the attainment of higher CONQUAS scores. Although it took a few years to see significant results, gradual improvement was observed as employees needed time to accustom to the new system. Overall, this can only be achieved when everyone in the organisation works closely together for the common good, to maintain an upward spiral of quality and productivity gains to propel the industry to a higher level. Company Z believes that the way forward is to change the mindsets across the entire organisation so that the renewed CONQUAS management framework is able to take effect.

7.7 Lessons Learnt

Based on the case study conducted, this research gathered that it is imperative for an organisation to fully understand where it wants to end up as well as appreciate where it currently stands and how it got there before implementing a new CONQUAS management framework. It is acknowledged that workmanship quality problems in construction stemmed from the earlier stages in a development and that if that area could be improved, the knock-on benefits could be substantial. From Company Z's experience, it is observed that the initial stages in the CONQUAS management process are crucial to the overall performance, and hence, stages prior to actual construction on site should be analysed and tackled first.

The above-mentioned descriptions have illustrated the past, present and future outlook of Company Z's main approach to manage the CONQUAS workflow. This account has illustrated the importance and usage of all the 33 CSFs, particularly the five major strategies which were brought up are found to have obtained relatively high mean scores in both the importance and usage findings. The findings have also revealed that the conceptual framework established in Chap. 4 is akin to the strategies adopted in Company Z, and hence, it will now be justified and detailed in the next chapter.

Reference

Corenet. (2012). CONQUAS projects. Retrieved December 17, 2012, from http://www.corenet.gov.sg/homeowners/listsql/default.asp?SortBy=Q29udHJhY3Rvcg==&category=UFJJVkFURSBIT1VTSU5HIChBTEwp

Chapter 8
CONQUAS Management Framework

8.1 Model for CONQUAS Management

With the literature review and research methodology undertaken, the conceptual framework developed at the end of Chap. 4 was shown to be constructive for use by contractors to better determine how the 33 CSFs contribute to the attainment of high CONQUAS scores. The CONQUAS management framework demonstrates how contractors can develop and value-add to their CONQUAS planning before actual construction to realise the goal of achieving a high CONQUAS score. Hence, the following will explain how the framework was developed.

This framework is structured into a Venn diagram (Fig. 8.1) to explain the concept of subset whereby all the elements of one set are contained within another set. In this study, "Stage B—Construction Planning" is also in "Stage A—Input Planning", and hence, Stage B is a subset of Stage A which is denoted as $B \subseteq A$ or {Construction Planning} $\subseteq$ {Input Planning}. The universal set (U) contains quality management practices which are not specific to CONQUAS management and are not within the scope of this study. The CSFs are grouped and linked together according to the results of the correlation analysis as far as possible.

This framework simulates from a basic construction process that transforms the input to output. From one lower point to another higher point in the framework, some new activities or techniques are introduced which serve to improve the workmanship quality of construction. The inspiration of this framework comes from the ISO 9001 Plan–Do–Check–Act (PDCA) process model which is widely accepted and recognised. The inputs are developed based upon the design and construction processes to assure quality in the design, in the materials and components used and in the site procedures used for communicating performance standards, and an effective monitoring, control and feedback mechanism set-up for the actual construction to reinforce workmanship quality in the final built product. Hence, the following sections will seek to explain the CONQUAS management framework based on the above analysis.

S. P. Low and J. Ong, *Project Quality Management*, DOI: 10.1007/978-981-287-074-2_8,

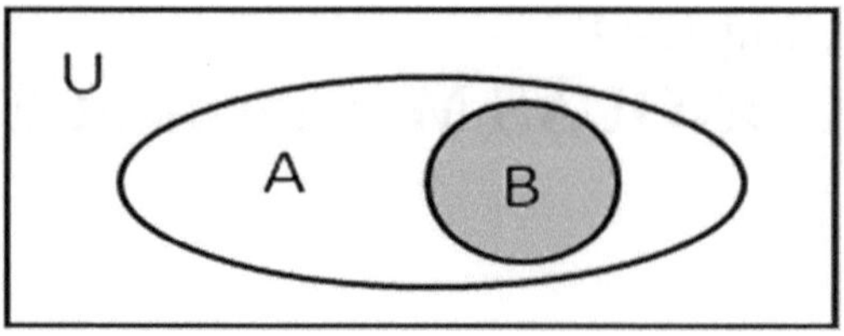

Fig. 8.1 Venn diagram—stage B is a subset of stage A. *Source* Schumacher (2012)

8.2 Stage A: Input Planning

8.2.1 Stage A1: Human Resource Management

Firstly, the start of this CONQUAS management framework requires the active involvement of the top management throughout the whole framework and they must really understand the entire system and its benefits. The top management will provide the resources for the entire model to proceed smoothly and address any chronic problems of an urgent nature as well as provide for reward and recognition to the QMU to stimulate better workmanship quality as compared to their past projects. As the top management kicks off the CONQUAS management framework, there will be some intrinsic and extrinsic factors which play a part in influencing the model from advancing to its fullest potential. To begin with, the main contractor, as the chief party in command of the CONQUAS management workflow, will have to make sure that their employees are well trained to form the QMU and carry out the CONQUAS management activities effectively. Upon benchmarking, this framework can be further developed according to the needs of the project, based on its goals, objectives of the implementation and the nature of the organisation. A strong commitment to workmanship quality is also required from the various parties involved as this model progresses and continues to the following stages to guide the CONQUAS management strategy.

The suggested structure of a QMU follows the structure of a QCC where groups of staff form QCCs and each member of the first level is to be the leader of the second level and so on. The function of QCC changes during the implementation of the CONQUAS framework from developing the model to monitoring and implementing the system. Any difficulty should be raised and resolved in the circle so that a practical framework particularly for the project can be achieved. Each QCC will assist in conveying the top management's mission to the lower level and provides the feedback from the lower to the upper level to involve all staff in the process of executing this model. Staff can be divided according to department and job function, and they are assigned with specific tasks, and extra working load to anyone of them is avoided. The size and the number of levels depend on the size and complexity of the project. Higher level circle has to meet and discuss more frequent than the lower circle and at the initial stage, every circle should meet more regularly to create an effective communication link between the top, middle and lower level. Each circle consists of staff from the same area, and they are accountable to perform the CONQUAS management activities to achieve good workmanship quality standard.

Therefore, this shows that workmanship quality performance and HRM (rated as the most important aspect in the survey conducted) are bound tightly together which suggests that the way people are managed is a very important determinant of success in achieving high CONQUAS scores. Stage A1 (Fig. 8.2) explains how success is determined by a contractor's skill in implementing the seven HRM CSFs to give it a measurable advantage and how the top management's commitment to developing its people's abilities and skills should be an obligation at all levels in the QMU.

8.2.2 Stage A2: Subcontracts Management

Secondly, it is noted that the main contractor normally subcontracts a major portion of work to the subcontractor and procure materials from suppliers who report directly to the main contractor. Subcontractors and suppliers do the actual physical construction and have a direct influence over the workmanship quality. Hence, they have to be pre-qualified through a stringent process based on good relationship and good track record for on-time completion as well as proven records to produce good quality end product. Subcontractors and suppliers should have their own quality programs which start from the material selection to its installation on site. Although competitive pricing is important to decide on the subcontractor to work with, those who have proven records in delivering quality products will be given bonus points in the selection process. This is because only then can their pledge to achieve a specific CONQUAS score be assured.

Furthermore, the main contractor has to avoid having too many different subcontractors working on site so as to involve each party with a larger proportion of work and by having a greater interest, be more committed about the achievement of high CONQUAS scores. This can be encouraged by awarding incentives and imposing penalties to those who do not follow the CONQUAS management workflow and as a result, affected the CONQUAS scores negatively. Lastly, the contractual period and contract sum have to be realistic as well so that they are able to complete the project on time and also within an appropriate budget to produce good workmanship quality.

Stage A2 (Fig. 8.3) shows how the nine SCM CSFs play a part in leading to the achievement of high CONQUAS scores. Examples of the evaluation and assessment criteria of the subcontractors and suppliers are also presented, and it should be noted that Stage A2 is linked to Stage A1 as people are the ones who will be managing the subcontracts.

8.2.3 Stage A3: Schedule Management

Prior to the commencement of works, the construction schedule will have to be arranged to indicate the timing and sequence of construction activities and the overall completion time. It is thus crucial to employ appropriate scheduling and

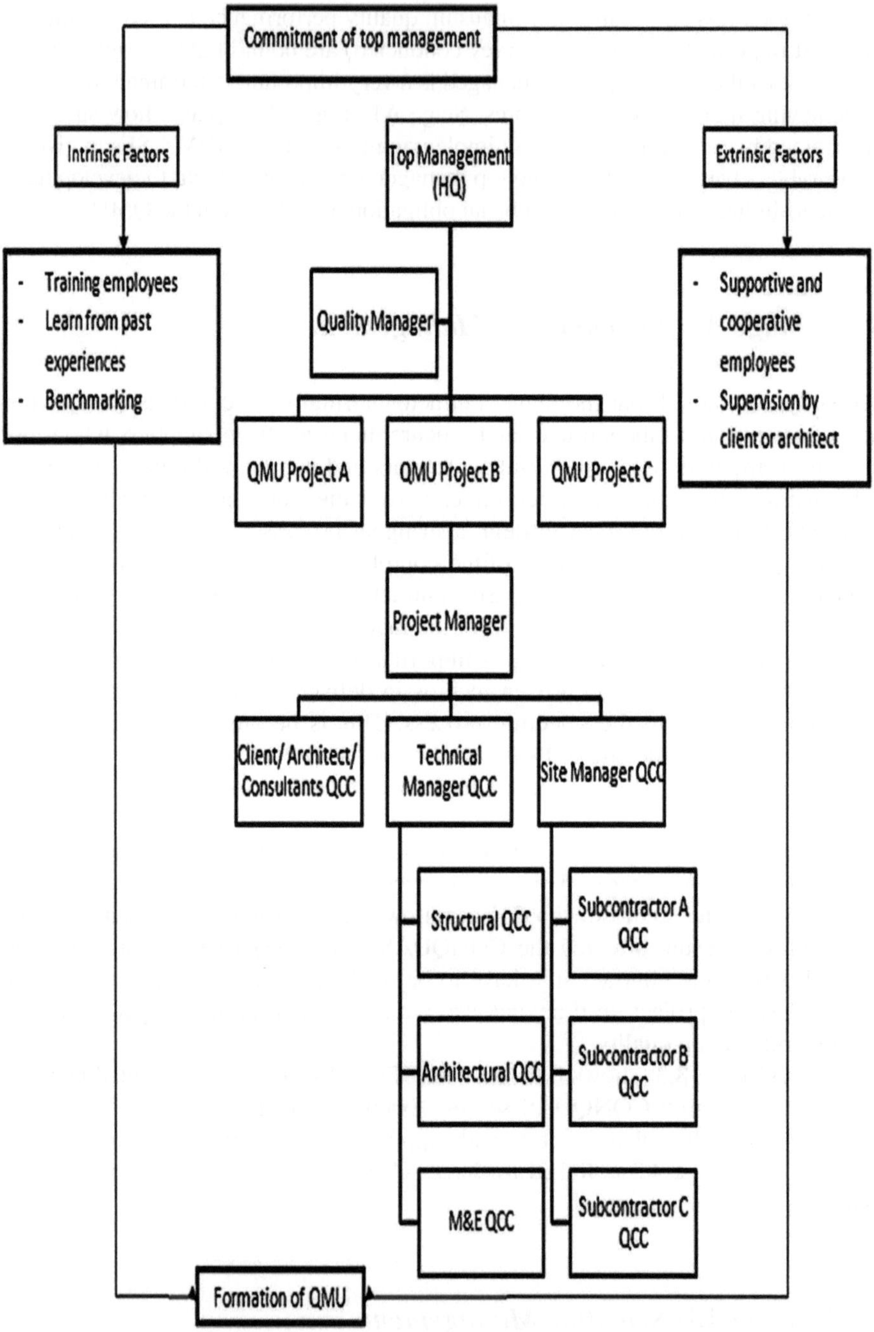

Fig. 8.2 Stage A1 human resource management model

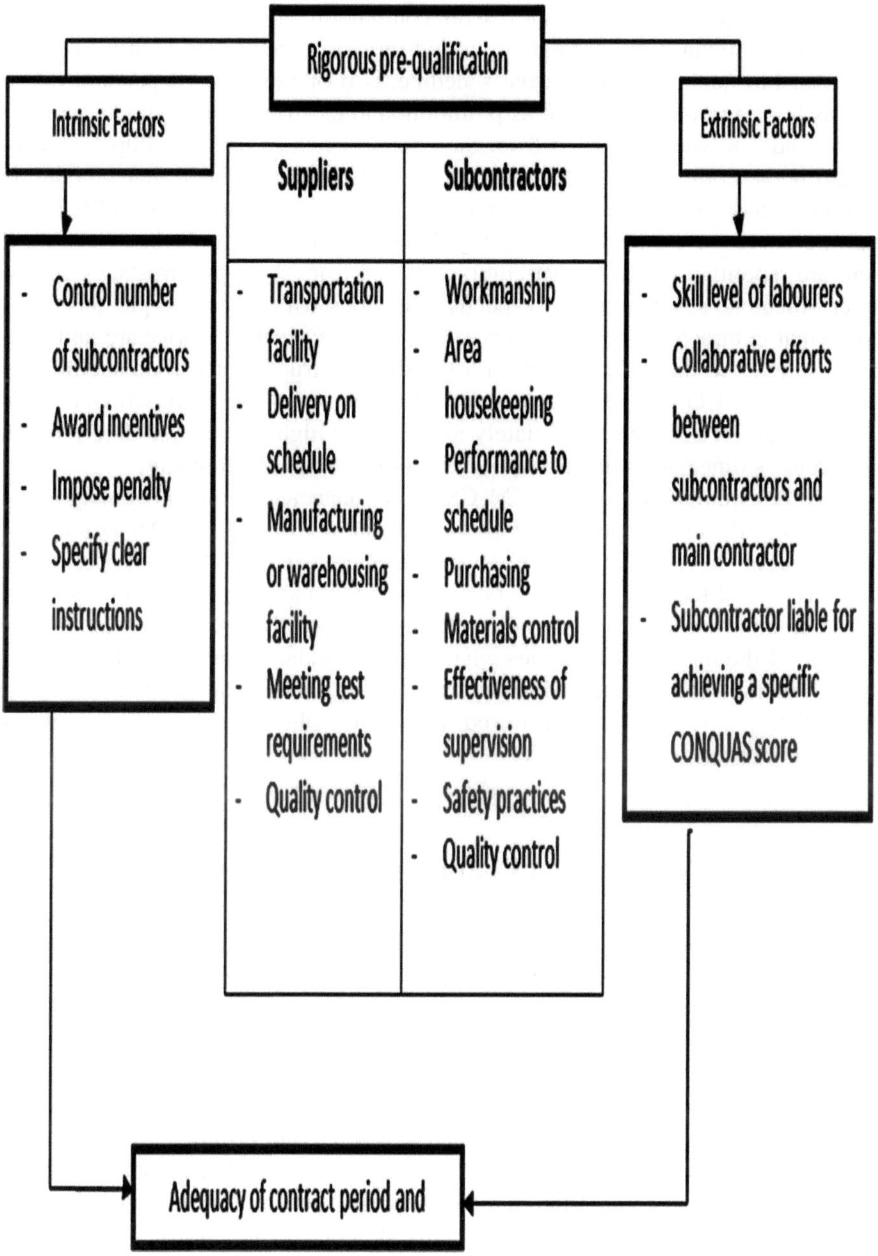

Fig. 8.3 Stage A2 subcontracts management model

control techniques to achieve successful completion of the project within time, budget and without disregarding safety and most importantly, to assure excellent workmanship quality. Based on the schedule, two or more interdependent trades such as flooring or tiling as well as plumbing and electrical works should work as a team and ask what they need from the others to reduce damage, improve quality and boost craftsmanship (Low 1998). It is also recommended that scheduling for phase construction is effective to allow contractors to reap the benefits of the learning curve on workmanship quality. However, this may only be workable if there are repetitive works so that contractors can learn from past mistakes and do better. Adequate samples in the respective locations will have to be pre-scheduled too so that contractors will know where the areas of greater priority are and allocate resources to them accordingly. Lastly, the construction schedule should also set the date for CONQUAS assessment so that once each section of works is completed, it can be assessed immediately as there is a high chance that deterioration will occur as time passes due to exposure to the external environment.

Stage A3, comprising of four SM CSFs, is connected to the above stages and completes part of the input planning stage as shown in Fig. 8.4. An example of how schedule planning for CONQUAS management can be done is also presented. Stage A1–A3 of the input planning is a set of requirements that should be planned at the beginning of the project to help achieve workmanship quality throughout the lifecycle of the project. It identifies what quality standards are required and determines how they can be satisfied. Essentially, input planning serves as a control to ensure that quality is not compromised. If it is not followed, it may cause rework and a delay in the project schedule.

8.3 Stage B: Construction Planning

Besides the human resources (HRM), subcontracts (SCM) and schedule arrangement (SM), the input planning stage is also extended to cover the material (MM) and construction aspect (CM) whereby the selection of building materials and construction method will have to be discussed to identify the most suitable and practical way to carry out the works in Stage B. The QMU should be trained to follow these systematic procedures in Stage B with the objective to ensure that the incoming materials are purchased, manufactured, delivered and installed when required, in the right quantities, right time and correct place. Essentially, construction planning is a set of inter-related internal sub-processes, each of which contributes to the overall process to deliver the final built product to the utmost workmanship quality standard.

It is suggested that the project manager will commence the construction planning with the process of selecting suitable materials which covers the QA obligation of the supplier. Any action affecting the workmanship quality performance should also be included. Next, review of shop drawings is essential to help the

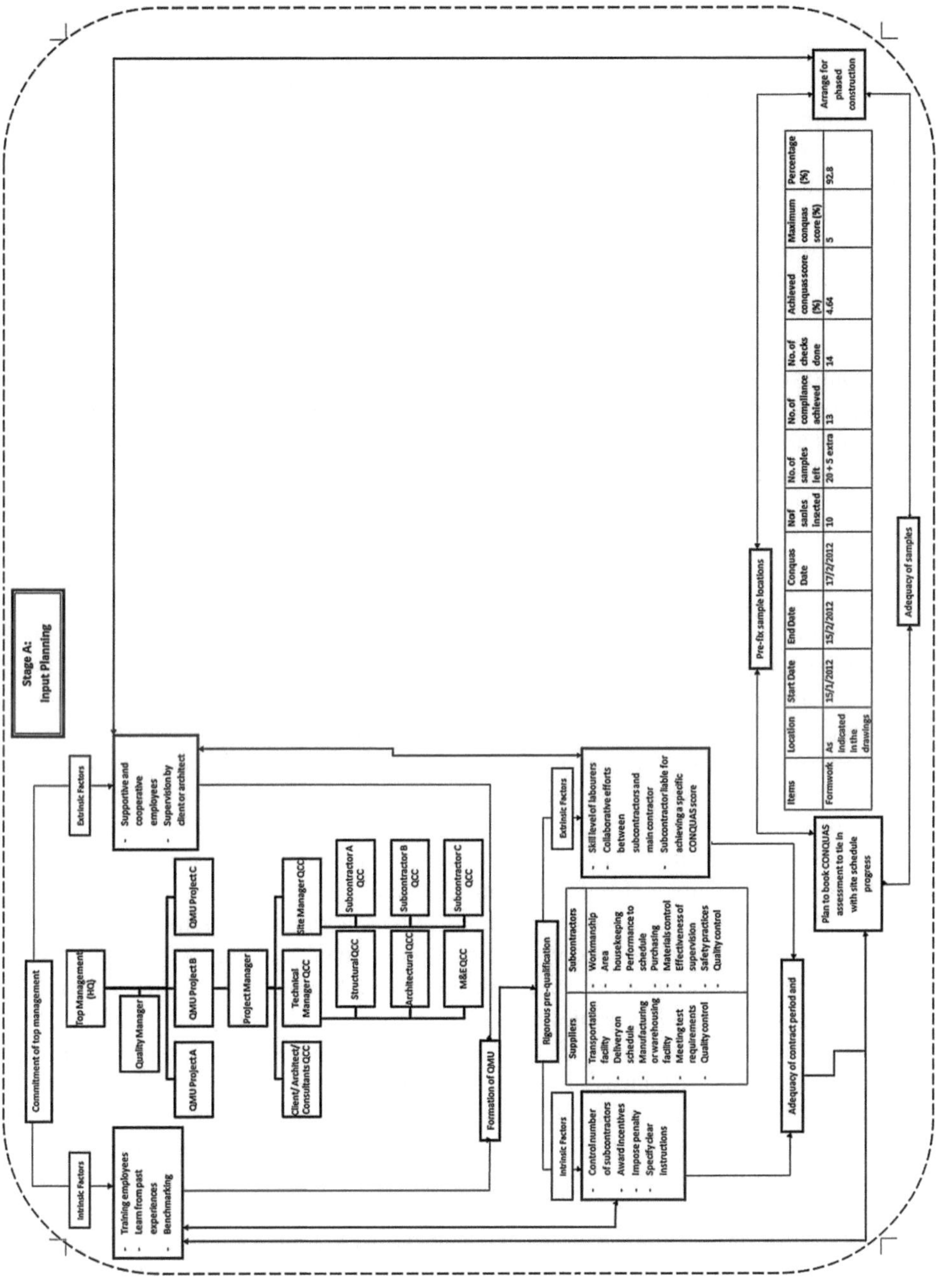

Fig. 8.4 Connection of stage A3 schedule management to stage A1 and A2

construction team mentally visualise the construction process before actual work commences. Details of the shop drawings will also have to be checked to ensure that it satisfies workmanship quality requirements. In addition, updated information is vital to make sure that the construction elements are constructed according to the right and latest drawings and information. Submittals are the basis for determining work conformance and adherence to design intent (Demkin 2008). This is vital because if project submittals were not required, it would be difficult to determine whether the construction conforms to the design intent until the work is in place.

Next, once materials are confirmed and delivered, it is critical to conduct an inspection as a pre-construction requirement to ensure that the material is acceptable and conforms to specifications. This inspection process also handles the material storage upon delivery with the primary aim of preventing and detecting any non-conformity before actual construction to ensure better quality of the final built product. At the same time, the materials should also comply with recognised standards and for some components, these have to be tested by an ITA before proceeding. Using this model, demonstration of a subcontractor's and supplier's capability to design, supply and construct the product primarily aimed at preventing non-conformity at all stages from design through to servicing can be affirmed as well.

The construction plan has to cater for conducting field demonstration and constructing mock-ups as far as possible for all trades to assure that the ultimate workmanship quality will be good. If the field demonstration and mock-ups do not pass the quality requirements, the process goes back to the previous stages depending on what the problem is. Only if the field demonstration and mock-ups meet the quality requirements can actual construction works be approved and constructed with protection thereafter before moving on to the following stages. Attention has to be given to protect the constructed components because of the need to avoid any form of damages while other works are still on-going to prevent the final workmanship quality from being affected.

So far, the above construction planning inputs adopt a predictive or preventive approach, but once actual construction is completed, the only way to improve the workmanship quality is a corrective approach. Although it would be perfect if workmanship quality is achieved without utilising any corrective actions and simply by assuring that the predictive and preventive processes are carried out in the right way, such instances are rare. It is noted that occurrence of defects and reworks will still be lower upon executing the previous stages and is critical to achieve zero defects (Ong 1997). Hence, the corrective approach is still essential to improve the probability of success in attaining excellent workmanship quality and a high CONQUAS score.

Preparatory inspections at every stage of the works have to be used as a tool to determine whether the component should be accepted or rejected based on the specified requirements. If works to some location are not carried out according to the requirements and is allowed to proceed with the following works, then it will be more expensive to rectify later. For example, if the vibration for

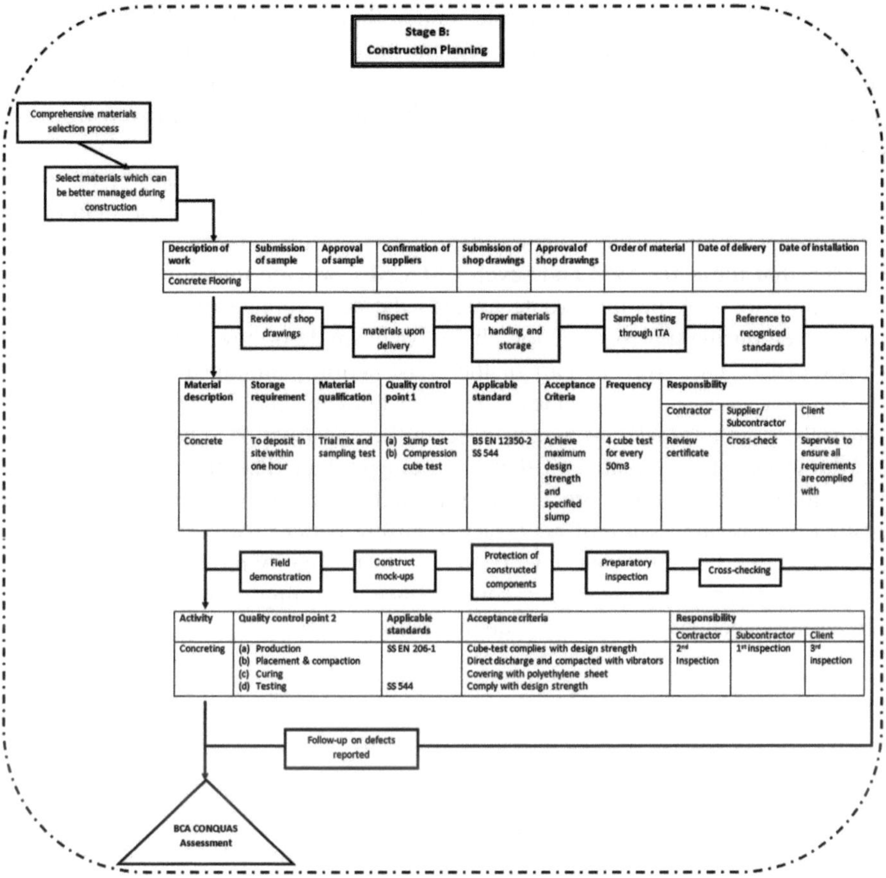

Fig. 8.5 Stage B construction planning model

compaction during concreting is not properly done, honeycombing would be found when the formworks are removed. Hence, preparatory inspection should be a standard pre-inspection procedure before the final inspection and ideally, cross-checking by a third party is also recommended for further scrutiny to maximise the workmanship quality performance and get ready for the actual CONQUAS assessment.

When defects or any components that do not perform to the accepted level are found, it should be placed as a top priority to be rectified immediately to prevent the non-conformity from worsening. The cause of the problem has to be identified as well, with actions taken to prevent recurrence, for example, by changing relevant practices to improve subsequent workmanship quality. This is because any defect found and rectified at the early stage is less costly and time effective to align the internal construction sub-processes to an acceptable level. Moreover, as workmanship quality problems frequently originate from earlier minor ones, any unaccepted works will have to be rectified before the later activities can continue. For example, concreting works are to be inspected to ensure that elements such as reinforcement, formwork and so on are in the right quantities and placed in the correct way. Through this method, serious defects will be reduced and the overall workmanship quality can be improved from the earlier stage.

Stage B is a subset of Stage A as it should also be done during the input planning stage before the construction phase begins as well as throughout the entire construction period. Regardless of the approach taken, it is shown in Fig. 8.5 that these six MM and seven CM CSFs provide many opportunities for contractors to work on their route towards the achievement of high CONQUAS scores even during the construction phase. An example of how concrete works can be planned is also given as part of the construction planning model. In particular, the first table shown provides a basis for knowing the time frame for quality materials to be approved so as not to delay the construction schedule and affecting the workmanship quality of the final built product. Therefore, each submittal needs to be reviewed and processed in a timely fashion, with sufficient time for dealing with resubmission, if necessary to ensure better quality performance. It is also noted that having a clear understanding of the quality requirements is critical in order to detect any form of non-conformance accordingly to further improve on and ensure excellent workmanship quality performance.

8.4 Output of the Framework

With the application of these 33 CSFs, workmanship quality of the construction works can be assured to a higher expectation level. The complete CONQUAS management framework will help contractors not only in assuring the quality of the final built product through systematic procedures, but it also helps in improving the reputation of the organisation who achieve high CONQUAS scores eventually. In return, this will also help the firm to secure more construction jobs in the future.

Besides high CONQUAS scores, continuous improvement is another fundamental benefit of the CONQUAS management framework which is also seen in Deming's cycle of PDCA, having a culture of never-ending improvement. With strong commitment to constantly work on identifying gaps in workmanship quality performance and developing the right strategies for closing them, the best construction practices can also be achieved. Feedback loops should be available to document and escalate any quality matters to the top management to share such knowledge on a company level. Opinion can be taken from the subcontractors and client as well. These feedbacks and information should be analysed carefully to seek further improvement by identifying elements of previous projects which have shown a high rate of quality non-conformance. This provides useful data for use in future CONQUAS projects to enhance the

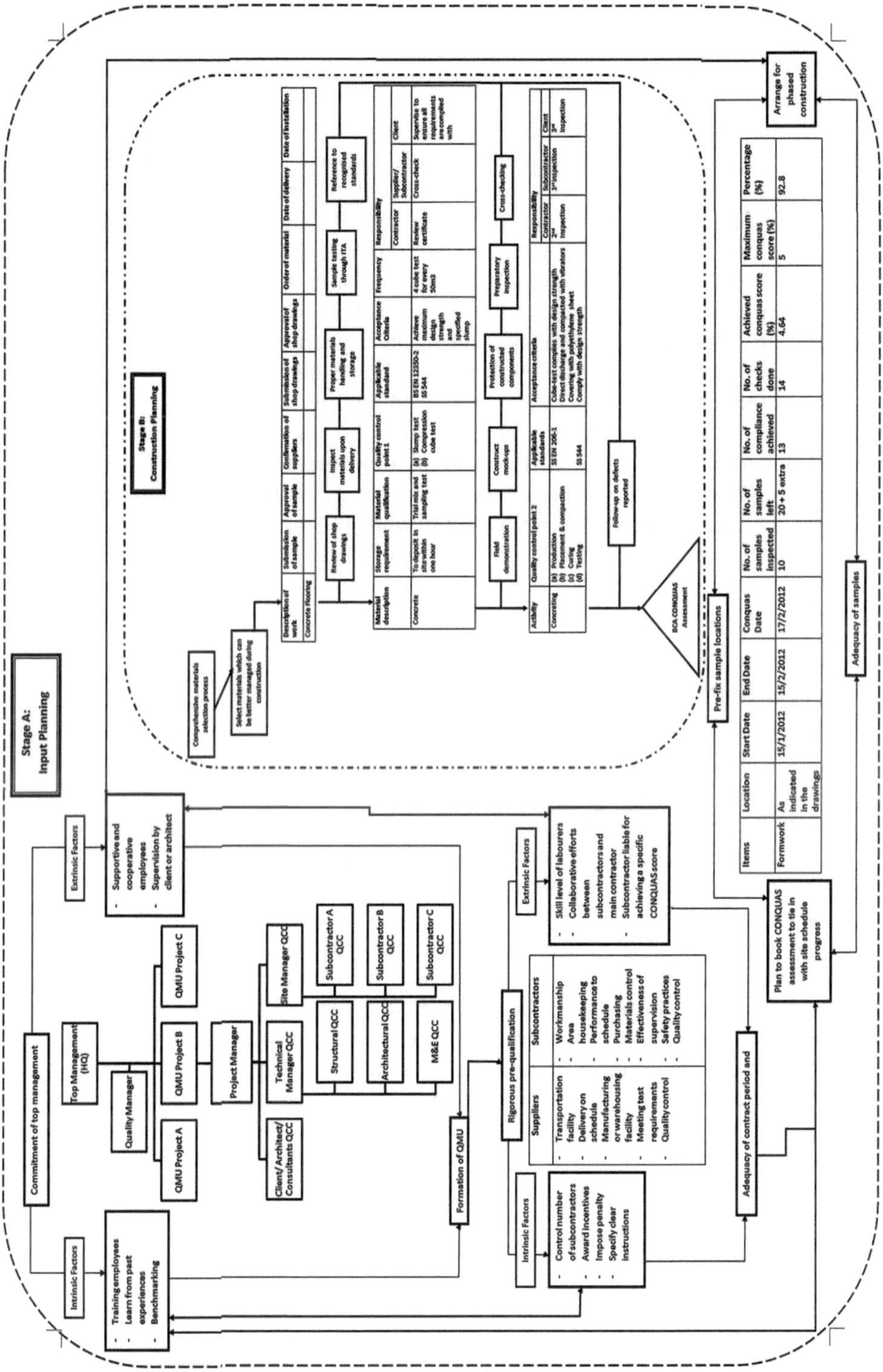

Fig. 8.6 CONQUAS management framework

CONQUAS management framework as well as avoid common pitfalls and abortive effort.

All in all, the above have explained the grounds for the entire company and industry as a whole to always be engaged in utilising the CONQUAS management framework (Fig. 8.6) to satisfy the goal of attaining high CONQUAS scores. In the process of moving from one stage to another stage, the management of workmanship quality will get better. In fact, this framework has adopted three quality management approaches, namely quality by design, quality by process control or assurance and quality by inspection (Ong 1997). It begins from intelligently designing quality into the whole construction process to the process-oriented views of controlling the quality characteristics of the components with a reference as well as effecting a fixed quality procedure throughout this process and finally, performing simple inspection of the finished components for elimination of defects and if deviations from the desired state are detected, taking remedial action to reinstate the targeted workmanship quality level. Dotted lines are used to enclose the framework to signify that they are subsets and actually originate from the universal set of quality management practices as described.

8.5 Application

To start with, this framework can first be developed at a company level and then a few projects can be selected as pilot projects for implementation. The projects selected are preferably of diverse nature but should consist of a well-trained and committed project team to execute the implementation process of the framework. This framework should be implemented during the early phases of the construction project for it to take effect. Furthermore, it is necessary to ensure support from the subcontractors, suppliers and various stakeholders of the project. The main contractor should also be ready with targets for workmanship quality improvements so that the subcontractors and suppliers do not view the entire framework as one-sided. Subsequently, an improvement in the CONQUAS score obtained will prompt the entire construction industry in Singapore to adopt a similar model. Lastly, it is also essential to make sure that staffs do not lose enthusiasm throughout when operating the entire framework.

8.6 Validation of the Framework

This framework is developed based on the literature study and survey findings. A quality assurance manager, an engineer and a quality practitioner were interviewed to evaluate the practicality of the CSFs and to get their comments on the feasibility of a CONQUAS management framework. They have been involved in various projects in the local construction industry for about 4–15 years and can be considered as fair representation because they are the ones who directly manage CONQUAS. Each interview took about one to two hours long because of the need to explain the objective of this research and the proposed model to them. All interviewees and respondents knew about the importance of the 33 CSFs beforehand but do not want to or do not know how to implement them effectively, or do not have the appropriate resources and expertise to do so or were constrained by the client's demands.

Generally, understanding of this framework was rather straightforward but it still has to be tested in the local construction industry for further validation. It is foreseen to be easy to apply which is a very important consideration as the main group who would be operating it is the QMU, and ease of use is needed to ensure better cooperation from other parties as well as site personnel. It is also flexible to mesh in with different circumstances and meaningful to the achievement of high CONQUAS scores.

Although this framework was commented by the interviewees to be practical and useful for full-scale implementation, it requires time to educate and train the people involved to implement such a new structure effectively. Moreover, this framework is still not fully investigated for its practicality because the sample size was limited to the main contractors. Other stakeholders of the project have not yet expressed their concerns about their role in the framework. With that, the next chapter will wrap up this research study with the conclusion.

References

Demkin, J. A. (2008). *The architect's handbook of professional practice*. Hoboken, New Jersey: Wiley.

Low, S. P. (1998). Managing total service quality: a systemic view. *Managing Service Quality*, *8*(1), 34–45.

Ong, B. H. (1997). *Improving quality performance in construction using a total quality management model*. MEng thesis. Kuala Lumpur: University of Malaya.

Schumacher, P. (2012). *The Autopoiesis of Architecture: A New Agenda for Architecture*. Vol.2, London: Wiley.

Chapter 9
Conclusion

9.1 Discussion of Main Findings

On all counts, this research has addressed the workmanship quality problems of the structural, architectural and M&E components and introduced suitable CONQUAS management techniques by referencing to quality management concepts to introduce a CONQUAS management framework. All the three objectives identified in Chap. 1 were achieved in this research through various methods including literature review, questionnaire surveys, interviews and a case study. The data obtained were also analysed, and the following describes the conclusion reached.

9.1.1 Objective 1: To Examine the Development of CONQUAS

As it was revealed that much could be done to improve the CONQUAS scores, the detailed CONQUAS scheme and its trend were discussed in Chap. 2 where this research has highlighted some important problems.

(1) The need for CONQUAS to be reviewed periodically due to changes and development of construction technology so that the standard can sustain present trends in the industry.
(2) The fact that a greater number of structural samples are required for non-housing projects which led to better CONQUAS performance for housing projects in the structural component. Comparatively, the number of M&E samples required is low, and hence, the M&E component will be able to score better than structural and architectural works on average. As a result, a greater number of architectural samples are required for housing projects which led to better CONQUAS performance for non-housing projects in the architectural component.

S. P. Low and J. Ong, *Project Quality Management*, DOI: 10.1007/978-981-287-074-2_9,

(3) In addition, as the architectural component accounts for at least 55–65 % of the CONQUAS score depending on the category of building it belongs to the attainment of high CONQUAS scores is very well limited by this component.
(4) The need for actual construction erection to be planned to conform to the contractual requirements (downstream quality management activities) as well as receiving adequate specifications of the design drawings and material information (upstream quality management activities).

9.1.2 Objective 2: To Investigate the Application of QMS to Manage CONQUAS

The concept of a QMS was defined in Chap. 3 with emphasis on particular push and pull factors in carrying out the application process.

(1) Quality management will improve the level of satisfaction on workmanship quality (Ong 1997). However, it is misunderstood that the implementation of quality management will increase the total final cost and prolonged the project's duration while some are still unsure about the advantages of quality management. But in fact, the high cost of non-conformance can be minimised through proper quality management. This indicates that there can be great possibility of cost savings through quality planning.
(2) Resistance towards greater implementation of quality management in the construction industry stemmed from the lack of support from the clients such as limited time given to complete the project. Contractors often have to work overtime to speed up the construction which affected the quality of the construction works. Frequent changes to the works were also one of the main causes of projects being unable to attain excellent workmanship quality. In spite of that, it is vital that contractors remain driven to find better ways of implementing a formal QMS to improve their workmanship quality. Clients will also have to improve their ability to supervise the works as well as to control the construction quality.
(3) A quality management programme is essential for the implementation of successful QA practices and the application of QC. Even so, in order to effectively implement quality, people must understand what quality is and how it benefits everyone. It is the leadership responsibility to lay the foundation and support a quality culture within an organisation.
(4) Site personnel are generally poor in quality management techniques, communication and project planning skills. Hence, it is important for the top management to devote more attention to emphasise on doing things right the first time to them in order to effectively apply quality management practices to CONQUAS and for results to be achieved on site.

9.1.3 Objective 3: To Identify CSFs for Achieving High CONQUAS Scores

This objective is achieved by the literature search as described in Chap. 4 and analysed in Chaps. 5, 6, 7 and 8. Based on the above facts, it was gathered that quality management techniques (universal set) are in fact applicable to CONQUAS management (subset). Hence, this research has taken the approach of introducing CSFs which are significant in helping to achieve high CONQUAS scores and activated the CSFs into a CONQUAS management framework. Besides that, examples were also given to illustrate the implementation process in the framework. Although these CSFs are known to them, the extent of adoption is still not high and this may be due to the lack of awareness of their importance as well as limited financial capacity and experienced manpower to implement them to attain high CONQUAS scores. It is also important to note that such quality movement has to be sustainable so as to bring about the essence of the evolution of CONQUAS, to produce better workmanship quality together with changing trends in the industry.

The inputs of great significance in the CONQUAS management framework are highlighted as follows:

- The CONQUAS management framework has to be initiated from the top management level, and all stakeholders have to play a part in the implementation process under close supervision of the top management as well as the client through QCC. This serves to guide the lower levels to execute the CONQUAS management activities effectively.
- The lack of knowledgeable and skilled staff which led to unacceptable construction quality is a sign that relevant trainings are essential to equip all personnel with the respective technical know-how to manage the construction quality efficiently. Training is very important to prevent poor utilisation of the framework which may lead to inefficient methods of managing the quality of a project. This will enable them to be more involved and concerned of the workmanship quality rather than simply focusing on the progress of the project and fully relying on aftermath inspection process.
- It is also noted that factors under the construction planning stage ought to be started as soon as a QMU is formed to prevent workmanship quality problems from surfacing before actual construction and to correct workmanship quality defects before CONQUAS assessment.

Therefore, it is viewed that the proposed CONQUAS management framework will provide contractors with a step-by-step guide from the very beginning, from no quality management at all towards the achievement of total quality construction. This framework defines a construction process as the transformation of a set of inputs (information, material, equipment and labour) into output (good workmanship quality). In the process, works will undergo inspection and is accepted or rejected and rectified before delivery to the next party involved. Besides that, all works are to be carried out according to the correct way accumulated from past experiences. It is also essential to constantly identify gaps in quality performance and develop the right strategies for closing them through continuously implementing the best construction practices in order to close the gap between the high- and low-performing contractors.

9.2 Validation of Hypotheses

To sum up, it was found that the following seven hypotheses were accepted:

(1) *Alternate Hypothesis,* $H_{1.1}$: *The more important the CSF is, the greater the influence it has on the CONQUAS score than CSFs which are not as important.*

Null Hypothesis $H_{1.0}$ was rejected, indicating that there are different important elements (Features and Basic Attributes) that allow certain firms to achieve higher CONQUAS scores than the rest. Hence, these 23 "Features" which are much more critical than the remaining ten "Basic Attributes" should be given additional attention in order to ensure that these 23 CSFs are put into action to materialise the goal of achieving a high CONQUAS score.

(2) *Alternate Hypothesis,* $H_{2.1}$: *The higher the adoption of the CSFs, the greater the resultant CONQUAS score will be than when the CSFs are not adopted.*

Null Hypothesis $H_{2.0}$ was also rejected, meaning that the different usage level of the CSFs in the CONQUAS management workflow allows certain firms to achieve higher CONQUAS scores than the rest. Hence, those five CSFs with low-usage level and are rated as "Features" should be taken note of as disregarding them will result in a greater effect on the CONQUAS scores.

(3) *Null Hypothesis,* $H_{3.0}$: *The extent of variation of the usage level of the CSFs cannot be attributable to the variation in its importance rating.*

Null Hypothesis $H_{3.0}$ was not rejected, meaning that the higher importance level of the CSFs in the CONQUAS management workflow does not necessarily result in higher usage level of the CSFs. The main reason for this phenomenon is due to the limited resources that contractors have which compete with other needs of the project such as the time, manpower and cost required to take the extra mile to implement these CSFs to realise the benefits of good workmanship quality which may not be successful after all the effort put in.

(4) *Alternate Hypothesis,* $H_{4.1}$: *The importance level of each of the CSFs is correlated to the importance level of other CSFs.*

Null Hypothesis $H_{4.0}$ was rejected, meaning that the higher importance level of a particular CSF will lead to a higher importance level of the respective CSFs and vice versa. This has been validated in view of the understanding that a moderate linkage can be drawn between them, especially those under the same category, which justify the design of the proposed CONQUAS management framework, grouping the CSFs under the same category together.

(5) *Alternate Hypothesis,* $H_{5.1}$: *The extent of adoption of each of the CSFs is correlated to the extent of adoption of other CSFs.*

Null Hypothesis $H_{5.0}$ was also rejected, meaning that the higher usage level of a particular CSF will lead to a higher usage level of the other respective CSFs and vice versa. This suggests that the CSFs are actually inter-related and largely dependent on each other in order to assure the successful progress of the entire CONQUAS workflow.

(6) *Null Hypothesis,* $H_{6.0}$: *The responses received from* A1 *contractors do not differ from* A2 *contractors with regard to the importance level of the CSFs.*

Null Hypothesis $H_{6.0}$ was not rejected, meaning that there is largely no distinction in the responses received from A1 to A2 contractors with regard to the perceived level of importance of the CSFs. It could be reasoned that A1 contractors can undertaken projects up to S$150 million, while A2 contractors' tendering limit is only S$65 million, and hence, A1 contractors may also be as financially constrained as the A2 contractors and may be in the same situation which A2 contractors face when managing the CONQUAS activities.

(7) *Null Hypothesis,* $H_{7.0}$: *The responses received from* A1 *contractors do not differ from* A2 *contractors with regard to the extent of adoption of the CSFs.*

Just as what has been explained in $H_{6.0}$, Null Hypothesis $H_{7.0}$ was also not rejected, signifying that there is largely no difference in the responses received from A1 to A2 contractors with regard to the degree of usage of the CSFs.

9.3 Implications to the Industry

Prior to this study, there was little research dedicated to CONQUAS management in the local construction industry, and none that studied on the importance relationship and extent of adoption relationship. This study will help industry professionals understand the importance of the 33 CSFs in construction and enhance their workmanship quality through implementation of a CONQUAS management framework, comprising of inputs from human resource, subcontracts, schedule, material and construction planning.

Although this framework for use by contractors downstream has been developed, no model which associates how exactly designers upstream should act has been conceived. Nevertheless, this framework will still be able to allow better understanding of how the 33 CSFs can lead to higher CONQUAS scores and to aid in the identification of potential key factors which designers can contribute as well and eventually leading to overall higher workmanship quality standard at the national level.

For this to happen, BCA has to play a role in raising awareness of the benefits of using this framework. Moreover, the employment of these CSFs as CONQUAS management practices is not only a solution to workmanship quality problems. Many problems such as the safety of the projects whether under construction or after completion, late delivery, higher final cost and accidents during construction can be minimised as well. When clients are aware of the CONQUAS management workflow, they should then be able to understand and work together with the project team to enhance the CONQUAS score and yet not downgrade on their quality demands.

However, the output of a high CONQUAS score is only based on meeting the specification requirements as indicated by the consultants and may not truly represent the workmanship quality requirements of the client. Therefore, it is important to get the involvement of the client in the QMU to participate in the process of monitoring the CONQUAS management workflow to assure that the workmanship quality stipulated will also satisfy their quality expectations. Although this CONQUAS management framework is still relatively new, the concepts are derived from recognised quality management techniques (universal set), and hence, it is believed that with active implementation, this framework will become an important ubiquitous quality manual for contractors in time to come.

9.4 New CONQUAS 2012 Requirements

However, it should be noted that as the latest eighth edition of CONQUAS was released on 31st October 2012 which came about mid-way during this study, hence the new requirements were generally not included in this study. Similarly, this eighth edition has undergone numerous consultations with industry partners and practitioners, data collection and analysis to establish a framework to enable the CONQUAS standard to keep pace with current practices in the industry. In fact, the main intention of the CONQUAS eighth edition is in line with the scope of this study, which is to promote the achievement of higher CONQUAS scores.

The eighth edition emphasises on the promotion of design and materials, which support both quality and productivity at the same time, to the achievement of higher CONQUAS scores. Bonus points are now given to projects to encourage the use of designs and materials which lead to higher quality and productivity. Thus, there will be better quality which means higher CONQUAS score for projects with buildable design and better material choices. In addition, a maximum of 1.5 architectural bonus points will be awarded to projects subjected to the QM scheme based on the average QM unit score and overall water ponding test passing rate achieved in the QM calculation. Furthermore, a full CONQUAS point will be awarded for employing a certified CONQUAS/QM Personnel. This will help to improve the CONQUAS score with the expertise and inputs of a CONUQAS-certified manager or supervisor.

Next, the eighth edition intends to optimise the value of the CONQUAS standard. Although higher CONQUAS scores generally reflect better workmanship, there is a diminishing return effect when the scores reach the higher limits. More effort is required to achieve the higher score without a marked corresponding improvement in observed quality. Besides, some developers and builders compete to achieve the highest CONQUAS score possible often at the expense of increased manpower. To prevent this, the CONQUAS score is now capped at 95. Any project scoring 95 and above will be rated as "CONQUAS STAR", and its score will not be published in BCA's website. This will deter the setting of high target scores above 95 points which will result in inefficient use of resources and thus reducing productivity (BCA 2012a, b).

In addition, it is understood that the majority of defects complaints concern finishing works due to the fact that architectural trades are more visible to end-users as compared with structural works. Hence, in order to align CONQUAS standards to end-users' expectations, the eighth edition has increased the scoring weightage by 5 % on architectural works and reduced by a similar percentage for structural works. Furthermore, a study and data analysis revealed that end-users are more concerned with floor and wall elements among the architectural trades, particularly cracks and damages and unevenness. Hence, the weightages of the quality standards for these elements have also been adjusted accordingly to closely better reflect end-users expectations.

Moreover, developers and builders can now monitor the workmanship quality of their projects with a new mobile application introduced by the BCA. This CONQUAS application is part of BCA's efforts to make CONQUAS or QM information easily accessible to industry professionals and even homeowners who are "on the move". There are three main unique features in this application which are not available online. Firstly, tips are provided to prompt users on defects that are frequently made by contractors. Secondly, users can monitor the workmanship quality of their on-going projects using the calculator to immediately generate a score based on different editions of CONQUAS. This check can be done before the actual CONQUAS or QM assessment by BCA. Thirdly, the CONQUAS checklists will provide users with a ready guide to prepare for CONQUAS assessment and improve the contractor's productivity.

All in all, such continual improvement made to CONQUAS drives the construction industry towards better built quality and productivity using fewer resources—time, cost and manpower to achieve improved workmanship quality together with productivity.

9.5 Limitations and Challenges of Study

Besides the exclusion of the CONQUAS eighth edition as a limitation, the results of this study may only be considered as exploratory since it was not implemented in reality and hence could not be truly accounted for and any anomaly observed may be explained by the following:

- In the construction industry, there are many dynamic variables which will affect a construction project. Thus, any changes to the workmanship quality or improvement to the CONQUAS score cannot be totally attributed to implementation of the CONQUAS management framework.

- The construction process involves various stages with various processes performed by different groups of people. Hence, it may be oversimplified that the subcontractors and clients and even internally where staff come from different disciplines will be easily convinced of the need to implement the 33 CSFs to achieving high CONQUAS scores.
- Although the CONQUAS management framework is adaptable according to specific project circumstances based on the quality objectives, it is limited as main contractors are often appointed only after the completion of all detailed design and this means that the main contractors do not have a chance to be involved or to contribute useful ideas to the design team to develop a high quality, productive and practical design concept to assure workmanship quality right from the design stage.
- The application of the CONQUAS management framework serves to satisfy specifications requirement to achieve good workmanship quality but what the client's actual quality demands are may not be satisfied. This only implies that contractors are able to meet the requirements established by the consultants and not from the end-users who probably should have the final and ultimate decision to say whether the expected workmanship quality standard has been fulfilled.
- As observed, some CSFs were deemed as unimportant in the survey, but literature findings have shown the positive impact of them. Hence, more research will have to be done to quantify the actual effect to the CONQUAS score. This may be due to the fact that the sample size of 38 respondents is still considered relatively small with a case study of only one company. Thus, the general trends captured in the survey and case study analysis may not be considered representative enough.

9.6 Recommendations for Future Research

An extensive and in-depth research can only be carried out by having full collaboration from industry partners and BCA to seek more information to develop more accurate results to be fully utilised and to promote workmanship quality awareness in the local construction industry. In addition, if time is not a restriction, this research will like to suggest the following for future study.

(1) The proposed CONQUAS management framework can be improved by appending a series of detail guidelines to further elaborate on the steps to be taken. A similar model can also be developed for use in consulting firms and client-based organisations to promote project-wide CONQUAS management.

(2) Besides improving the CONQUAS score, a thorough study on the benefits, practicality as well as difficulties in implementing the CONQUAS management framework can be determined to further enhance its potential. The research may also look into the time, cost and manpower changes in conjunction with the implementation of the CONQUAS management framework and probably the effect on each CSF, respectively. Benefits of this framework with regard to construction can also be included to assist industry professionals in justifying the adoption and practice of the proposed framework.

(3) A mechanism for benchmarking the CSFs in the local construction industry can be developed. This will be particularly useful to BCA as the centre of information in construction. Data such as the set of CSFs implemented and the actual CONQUAS scores obtained can be uploaded for the industry reference. Longitudinal case studies of a few projects of different building types can be conducted to test and verify the effectiveness of the CSFs to improve CONQUAS scores. This is to develop a measuring system to quantify the quality level of constructed works by capturing the CONQUAS scores obtained. For example, how exactly the application of pre-fixing CONQUAS sample locations affected the CONQUAS score.
(4) Currently, there is IQUAS which showcases the commonly encountered noncompliances. Hence, it is recommended that studies can be conducted to determine and quantify the exact reduction of CONQUAS scores with the particular nonconformities identified.

9.7 Concluding Remarks

To sum up, this research has showcased the means to the achievement of high CONQUAS scores. The proposed CONQUAS management framework is predicted to be of great value to contractors and will certainly have a great snowball effect on promoting the significance of having good workmanship quality. Even with such a defined QMS in place, contractors should still work towards continuous development, implementation and maintenance of the framework to keep pace with the changing quality trends in the construction industry.

References

Building and Construction Authority. (2012a). Achieve better quality and higher productivity through CONQUAS (8th ed.). Retrieved November 1, 2012 from http://www.bca.gov.sg/Professionals/IQUAS/conquas8.html

Building and Construction Authority. (2012b). Construction excellence award. Retrieved September 27, 2012 from http://www.bca.gov.sg/awards/constructionexcellence/construction_excellence_awards.html

Ong, B. H. (1997). *Improving quality performance in construction using a total quality management model*. MEng thesis. Kuala Lumpur: University of Malaya.

Appendix A: Survey Questionnaire

Survey on the critical success factors to achieve high CONQUAS scores

Dear Sir

I am a final year student from the National University of Singapore currently working on my final year project. The objective of this survey is to examine the factors to achieve high CONQUAS score and would sincerely appreciate your valued input. Please be assured that all data will be used for purely academic purpose and will remain strictly anonymous. Feel free to contact me for any queries.

Thank you very much for your participation.

Regards
Joy Ong

Section 1: General Information of Respondent

About how many years have you been in the construction industry? ______________

Please indicate (√) the Financial Grade (general building) of your company:

	A1
	A2
	B1
	B2
	Others: ______________________________

Please indicate (√) your designation:

	Project Manager/Director
	Technical Manager/Director
	Quality Assurance Manager/Director
	Others: ______________________________

S. P. Low and J. Ong, *Project Quality Management*, DOI: 10.1007/978-981-287-074-2,

Section 2: Importance Level of Factors

In your personal opinion, please rate the level of importance of these factors in influencing the CONQUAS score, where:

1 = Negligible
2 = Unimportant
3 = Neutral
4 = Important
5 = Extremely important

Human Resource Management	**Importance Rating**
Formation of a Quality Management Unit (QMU) with clear guidelines of each member's roles and responsibilities for every project	
Commitment of top management to advance the CONQUAS management performance	
Getting the support and cooperation of employees	
Training employees to understand CONQUAS requirements	
Supervision by the client/architect to monitor the CONQUAS management workflow of the main contractor	
Comparing company-wide CONQUAS performance track records	
Setting a benchmark to gauge and control the CONQUAS performance	
Subcontracts Management	**Importance Rating**
Controlling the number of subcontractors working on the project	
Having a rigorous pre-qualification process to select subcontractors and suppliers	
Ensuring the skill level of labourers	
Collaborative efforts between subcontractors and main contractor	
Adequacy of contract period and contract sum	
Identifying a specific CONQUAS score that the subcontractor is liable for achieving	
Awarding incentives to subcontractors if their targeted score is met	
Imposing a penalty to subcontractors if their targeted score is not achieved	
Giving clear instructions to the subcontractors on how to adhere to the CONQUAS requirements	
Schedule Management	**Importance Rating**
Pre-fixing of CONQUAS sample locations (applicable to structural components only)	
Adequacy of sample locations prepared in the event of unforeseen site conditions to serve as a backup	
Arranging for phased construction to reap learning curve on workmanship quality	
Booking of assessment schedule to tie in with site progress	
Materials Management	**Importance Rating**
Choose materials through a comparison process, considering their specifications and samples to be provided as well	

Select materials which can be better managed during construction	
Inspect materials upon delivery	
Proper materials handling and storage	
Protection of materials after completion of that portion of works	
Sample testing of materials through an Independent Testing Agency (ITA) to check for proper usage of materials	
Construction Management	**Importance Rating**
Ensuring that shop drawings are checked thoroughly before actual construction	
Constructing mock-ups to check for implications with other trades of works	
Field demonstration by labourers to showcase their understanding of the workmanship quality required	
Conducting preparatory inspection using template checklist at every stage of work activity	
Using recognised testing standards to check for non-compliance during inspection	
Cross-checking by another party (e.g. another supervisor, other subcontractors, client, architect, etc)	
Adherence to reporting and follow-up procedure of defects before CONQUAS assessment	

Section 3: Extent of Adoption of Factors

From the list of factors (below) that may influence the result of achieving high CONQUAS score, please put a tick (√) in the box if your firm is using such an approach or a cross (X) if your firm is not using such an approach, to manage CONQUAS in most of your projects.

Human Resource Management	
Formation of a Quality Management Unit (QMU) with clear guidelines of each member's roles and responsibilities for every project	
Commitment of top management to advance the CONQUAS management performance	
Getting the support and cooperation of employees	
Training employees to understand CONQUAS requirements	
Supervision by the client/architect to monitor the CONQUAS management workflow of the main contractor	
Comparing company-wide CONQUAS performance track records	
Setting a benchmark to gauge and control the CONQUAS performance	
Subcontracts Management	
Controlling the number of subcontractors working on the project	
Having a rigorous pre-qualification process to select subcontractors and suppliers	
Ensuring the skill level of labourers	
Collaborative efforts between subcontractors and main contractor	
Adequacy of contract period and contract sum	
Identifying a specific CONQUAS score that the subcontractor is liable for achieving	
Awarding incentives to subcontractors if their targeted score is met	
Imposing a penalty to subcontractors if their targeted score is not achieved	
Giving clear instructions to the subcontractors on how to adhere to the CONQUAS requirements	

Schedule Management	
Pre-fixing of CONQUAS sample locations (applicable to structural components only)	
Adequacy of sample locations prepared in the event of unforeseen site conditions to serve as a backup	
Arranging for phased construction to reap learning curve on workmanship quality	
Booking of assessment schedule to tie in with site progress	
Materials Management	
Choose materials through a comparison process, considering their specifications and samples to be provided as well	
Select materials which can be better managed during construction	
Inspect materials upon delivery	
Proper materials handling and storage	
Protection of materials after completion of that portion of works	
Sample testing of materials through an Independent Testing Agency (ITA) to check for proper usage of materials	
Construction Management	
Ensuring that shop drawings are checked thoroughly before actual construction	
Constructing mock-ups to check for implications with other trades of works	
Field demonstration by labourers to showcase their understanding of the workmanship quality required	
Conducting preparatory inspection using template checklist at every stage of work activity	
Using recognised testing standards to check for non-compliance during inspection	
Cross-checking by another party (e.g. another supervisor, other subcontractors, client, architect, etc)	
Adherence to reporting and follow-up procedure of defects before CONQUAS assessment	

THE END.
THANK YOU FOR PARTICIPATING.

Appendix B: Verbatim Report 1

This is an interview transcript with Mr A on the 20 November 2012, 10 a.m., at the interviewee's office. The interviewer shall be referred to as "J".

Begin Transcript:

J: Hi, as per our email conversation, my research topic is on the critical success factors for achieving high CONQUAS scores by contractors. Can you share with me on your company's approach to managing CONQUAS projects?

Mr A: We focus a lot on planning and ensuring that what is being planned can be realised on site. Our approach is more towards working hard on human resource management to achieve a high CONQUAS score as far as possible. Most of the time, the influencing factor for achieving high scores will be management support, awareness on importance of achieving target, site staff awareness, understanding the scoring system, penalties, regular site inspection and close monitoring of progress and scores. So, we work harder on these areas and at the same time will ensure that other factors affecting the CONQUAS scores are not neglected.

J: I noted that achieving high scores comprises of many factors and based on the mean importance rating gathered of these 33 critical success factors (pass paper to him), do you think that those higher ranking factors support your views on the strategies to managing CONQUAS?

Mr A: Hmmm…not really…I don't think that the higher mean importance rating means that contractors will input them as part of their strategies to achieve high CONQUAS scores. Yes, many a times, main contractors, sub-contractors and site staff do understand the importance, but when it boils down to practicing what they preach, it is a totally different mindset. So the policing part comes into play and how the QA guys do their inspection and how well the sub-contractors cooperate.

J: Do you have any comments with regards to these 33 factors in ensuring a high CONQUAS score?

Mr A: For the formation of a QMU with clear guidelines of roles and responsibilities for each member, it is also important that this QA or QC site team consists

S. P. Low and J. Ong, *Project Quality Management*, DOI: 10.1007/978-981-287-074-2,

of personnel who are responsible for fine tuning or touching up to prepare the area prior to inspection. Therefore, besides the inclusion of personnel from our company, we will also include the site supervisor of each subcontractor's firm in this QMU. Also, I think that there may be other external factors such as the experience of the site team in managing CONQUAS, the discipline of the people involved and maybe the requirements of the projects. But generally, if a company is able to achieve all these 33 CSFs successfully, I think that achieving a score above 90 is not a problem.

J: So, are you surprised with the mean importance rating of any of these factors, whether they contradict with your viewpoint?

Mr A: Somehow, the relatively low ranking of the adequacy of sample size prepared for inspection is not as important seems to be a common misconception among contractors. Although the minimum number of sample to be prepared is as indicated in the CONQUAS standards, based on my experience, the CONQUAS scores tend to be better if the entire block is ready for inspection. But in reality, this may be hard to achieve due to schedule issues and it is common that at most, only half of the block is ready and this may leave a poor impression to the assessors in general.

J: I have came out with an equation for contractors to use in estimating their CONQUAS scores based on the factors that they decide to implement in their project. (Pass paper to him)

Mr A: Oh…this is something new to me.

J: Basically, based on my statistical analysis, only 21 factors are found to be significant in predicting the CONQUAS scores which means that the result of the other factors happens only due to chance and hence, have to be rejected. However, this is only based on responses received from 38 respondents and just the 33 critical success factors identified. What do you think about it?

Mr A: I see. This is interesting. Let's do a test to this now (Takes out calculator and starts calculating). I have included those factors which we will definitely implement in all of our CONQUAS projects and the predicted score is a little lower than what we have actually obtained on average in the recent 3 years or so (Refer to equation below). Your equation may not be representative yet as there may be other external factors mentioned although the impact may be minimal. But to truly get results that represent industry-wide, it may be essential to gather responses from more contractors' firms; maybe at least 100 responses would be adequate to get an equation which is more credible.

*Predicted CONQUAS score = 75.117 + 1.014 (HRM1) + 1.164 (HRM4) + 0.429 (HRM6) + 1.197(HRM7) ~~**− 0.117 (SCM1)**~~ + 1.096 (SCM2) ~~+ 0.276 (SCM6)~~ ~~**− 0.134 (SCM7) − 2.461 (SCM8)**~~ ~~+ 0.468 (SCM9)~~ + 1.864 (SM1) + 0.262 (SM2) + 0.181 (SM4) + 3.064 (MM1) **− 7.314 (MM2)** + 1.817 (MM4) + 1.063 (MM5) ~~+ 4.436 (CM3)~~ + 4.432 (CM4) ~~+2.974 (CM5)~~ +1.271 (CM7) **= 86.7***

J: Then, what do you think of this proposed CONQUAS management framework that contractors can reference to during the course of the project to make sure that they have taken the appropriate steps? (Pass paper to him)

Mr A: I think the layout is quite clear and understandable. It is probably a good model to follow if the contractors do not already have a systematic way of handling the CONQUAS workflow activities.

J: I think that's all that I have to ask. Thank you for your time spent in helping me with my research.

Mr A: Don't worry. Feel free to contact me again should you need any help.

End of Transcript

Appendix C: Verbatim Report 2

This is an interview transcript with Ms B on the 6 December 2012, 8 a.m. at the interviewee's site office. The interviewer shall be referred to as "J".

Begin Transcript:

J: Hello, after receiving your feedback during the pilot survey, I have made some amendments and conducted a full-scale survey. So today, I am here to seek your opinion on the results of the survey.

Ms B: No problem. Sorry for such late reply anyway as we were really busy for the past few weeks tying up all the loose ends for our Phase 1A TOP. I hope I will be able to help you understand the CONQUAS system better, if not for your research; it can be just for your general knowledge.

J: No worries. Firstly, I noted that your company has a relatively poor record of CONQUAS score. Is this because you have only completed three CONQUAS projects thus far or…?

Ms B: Hmmm…the bulk of CONQUAS projects assessed are mainly residential projects and as far as I know, we do not tender for residential projects which may be because such projects are more stringent in the workmanship quality aspect and we are not confident of doing so. But we have been trying to improve our expertise in this area. In fact, this is our fourth CONQUAS project and at the moment, only structural and M&E assessments have been completed and results were pretty good.

J: Any particular reason for such improvement?

Ms B: Firstly, this is because we have to meet a minimum CONQUAS score as requested by the client as well as to meet the target score identified in our company's quality policy. Another reason may be because in this project, there is an experienced senior who has been involved in many CONQUAS projects in other firms. He is also a certified CONQUAS/QM Manager. He has given us a lot of tips in preventing defects from occurring by pre-empting us beforehand. He will also warn us to take extra care on areas, based on his experience, which have a high probability of non-compliance.

S. P. Low and J. Ong, *Project Quality Management*, DOI: 10.1007/978-981-287-074-2,

J: So, what is your company's approach to managing CONQUAS now?

Ms B: I won't say that CONQUAS management is consistent throughout all our projects as different projects are made up of different teams of people who may have their own way of doing things. But in this project, the QMU acts by setting out specific broad goals for the 'ultimate targets of quality'. The CONQUAS committee comprises mostly of experienced site personnel who understands the site conditions to ensure workmanship quality. Therefore, we are able to set up benchmarks that are practical and achievable on site.

J: Do you have any comments with regards to any of these 33 factors in ensuring a high CONQUAS score?

Ms B: Under human resource management, I feel that client or architect can only supervise or monitor but the main CONQUAS management workflow is usually being controlled by the main contractor. Then, arranging for phased construction to reap learning curve on workmanship quality is only workable if there is repetition in the buildings, that is, every storey is the same so that as one goes up, one learn from past mistakes and do better. With regards to the points about setting out clear guidelines, implementing a checklist for every inspection and stating clearly the requirements for each aspect or point of the guideline and probably some other factors, I feel that the main reason for that is to prevent any disputes and misunderstandings about the workmanship quality specified. ITA is also not implemented for the reason of ensuring high CONQUAS score but achieving high score is a consequence of that.

J: I have analysed the results of the survey and have ranked the mean importance rating accordingly as shown here (Pass paper to her). What is your opinion on the outcome of the result, in particular, the mean importance rating obtained?

Ms B: I feel that it would be great if contractors are able to implement all these 33 factors, but normally, such high mean importance rating factors tend to be much harder to achieve, in terms of having to spend more time and manpower in order to achieve. In that case, contractors may choose to spend their time and manpower on achieving more factors that have lower mean rating than just to spend all their time and manpower on a single high mean importance rating factor that at the end of the day, may still not be achievable despite all the time and manpower spent.

J: So, do you think that those higher ranking factors support your views on the strategies to managing CONQUAS?

Ms B: Some of the high mean importance rating factors are not difficult to achieve technically, just that more time, effort and proper planning is required for them to be achieved. Normally, contractors have a target CONQUAS score in mind and will then decide on the factors that are more achievable within their available resources, of which, most of the time will not be the high mean importance rating factors unless it is their forte. I guess these will then be the factors that are more important to them.

J: Are there any factors that I have missed out?

Ms B: Probably under material management, you may consider communication to the workers regarding the proper usage of the materials, especially for new products, as no matter how good the material may be, it may not perform as expected if it is being handled, stored or used in the wrong way.

J: I have came out with an equation for contractors to use in estimating their CONQUAS scores based on the factors that they decide to implement in their project (Pass paper to her to see the equation). Basically, based on my statistical analysis, only 21 factors are found to be significant in predicting the CONQUAS scores which means that the result of the other factors happens only due to chance and hence, have to be rejected. However, this is only based on responses received from 38 respondents and just the 33 critical success factors identified. What do you think about it?

Ms B: Erm…I am not very sure how you came out with this equation but in my opinion, I think that this may only be practical to firms like us who lack the expertise in CONQUAS management. Big firms or those who have been performing well may feel that it is redundant since they may already have a well-established system in place and knows what to do in order to meet their target scores, although not quantified in exact numbers.

J: Then, what do you think of this proposed CONQUAS management framework that contractors can reference to during the course of the project to make sure that they have taken the appropriate steps? (Pass paper to her)

Ms B: This is like a mind map? Perhaps this framework is more suitable for the subcontractors to look at so that they are clear about the main contractor's CONQUAS workflow. Subcontractors can also know what roles they have to play in the input planning and construction planning stage.

J: Any more comments that you would like to add?

Ms B: To a certain extent, having a good relationship with the BCA assessors can be beneficial. There are only a handful of assessors, maybe 15, so getting to know some of them are not difficult. However, the name of the assessors would not be known until the assessment date has been booked and confirmed. All that said, theoretically, BCA assessors are supposed to monitor the work qualities as a third party which has no grounds on either side in order for their reports to be fair.

J: I guess I have gotten the information that I need and certainly have a clearer picture with regards to how contractors manage the CONQUAS process. Thank you very much for taking time out by coming early to work to clarify my doubts.

Ms B: Welcome. Hope I make sense to you and all the best for your research.

End of Transcript

Appendix D: Verbatim Report 3

This is an interview transcript with Mr C on the 19 November 2012, 4 p.m. at the interviewee's office. The interviewer shall be referred to as "J".

Begin Transcript:

J: Hi, I presumed you have a rough idea of my final year project which is to find out the critical success factors for achieving high CONQUAS scores by contractors and today, I would like to obtain your views on this matter from the perspective of a quality practitioner.

Mr C: Yes, this is an interesting topic and is something that the industry is unaware of and needs to know too. I think nothing has been done so far to collate the critical success factors to meet high CONQUAS scores. Since you have attended the BCA-REDAS Quality and Productivity Seminar last month, you would have heard of the changes to the latest CONQUAS edition. In fact, the aim of the constant development to CONQUAS is to encourage contractors to achieve higher CONQUAS scores and not remain status quo.

J: I have analysed the results of my survey and have ranked the mean importance rating accordingly as shown here (Pass paper to him). What is your opinion on the mean importance rating obtained?

Mr C: I think that the factors are rather comprehensive and the factors under human resource management are the pre-requisites to meet the factors under other aspects like materials and construction management.

J: Do you have any comments with regards to any of these 33 factors in ensuring a high CONQUAS score?

Mr C: I feel that staff training on CONQUAS is very important. BCA has also recognised this and if I did not remember wrongly, BCA provides free trainings to four staffs involved in each project. Although this is only a one day workshop, I think that it is a basic requirement for all staffs who will be involved in duties related to workmanship quality. The latest CONQUAS edition has also provided bonus points for projects which employs certified CONQUAS supervisor or manager.

S. P. Low and J. Ong, *Project Quality Management*, DOI: 10.1007/978-981-287-074-2,

J: As CONQUAS training are also provided for developers and consultants and not just the builders, don't you think that they play an important role in ensuring a high CONQUAS score too?

Mr C: Hmmm…it's hard to say as some developers may not be that concerned about CONQUAS and totally allocate such risks and responsibilities to the contractors while others who want to ensure good workmanship quality will collaborate with the contractors to do so, especially in design-build projects. Of course, the best scenario would be that the upstream designing and downstream construction are both regulated to achieve high CONQUAS scores and not just the single responsibility of the construction process. But at this moment, in my opinion, the main deciding factor for upstream activities is still because of cost issues. A very little percentage of the selection is based on the fact that better workmanship quality can be obtained.

J: I see. Then, for CONQUAS assessment, it has to be done right the first time while the Quality Mark Scheme allows for re-scoring. Why is there such a disparity?

Mr C: For QM, the main reason is to encourage contractors to rectify any defects and deliver quality homes but for CONQUAS, the purpose is just to ascertain the workmanship quality level of the building constructed and defects need not be rectified and even if it is, re-scoring will not be done.

J: The achievement of high CONQUAS score will give an advantage to contractors tendering for government projects under the Price Quality Method. How is the effectiveness of this scheme so far?

Mr C: I feel that this doesn't seem to have a great impact yet unlike the Bonus Scheme for Construction Quality where feedbacks have been heard that contractors will go all out just to get the bonus payments to increase their profitability.

J: I have came out with an equation for contractors to use in estimating their CONQUAS scores based on the factors that they decide to implement in their project (Pass paper to him to see the equation used with an imaginary example). Basically, based on my statistical analysis, only 21 factors are found to be significant in predicting the CONQUAS scores which means that the result of the other factors happens only due to chance and hence, have to be rejected. However, this is only based on responses received from 38 respondents and just the 33 critical success factors identified. What do you think about it?

Mr C: Actually, I think that the potential of this equation concept is great but as what you said, it needs more credibility and reliability in order for something like this to be implemented industry-wide. This would be a useful tool if contractors are able to predict the CONQUAS scores just from which factors they decide to employ and which factors are not feasible as they may involve more time and resources, sacrificing on maybe other aspects like productivity, build ability and constructability.

J: Then, what do you think of this proposed CONQUAS management framework that contractors can reference to during the course of the project to make sure that they have taken the appropriate steps? (Pass paper to him)

Mr C: This is a good suggestion. I think this would be able to help contractors understand what they need to do to manage the entire CONQUAS process.

J: I see. I have no more questions. I really appreciate your help over these few months. Thank you.

Mr C: That's alright. All the best for your research.

End of Transcript

Appendix E: Statistically Significant Positive Correlation for Importance Rating

No.	Factor 1	Factor 2	Pearson correlation coefficient (r)	Sig. (two-tailed)
1	MM3	MM4	0.915	0.000
2	SM1	SM2	0.838	0.000
3	HRM4	SCM5	0.835	0.000
4	MM5	CM7	0.835	0.000
5	CM6	CM7	0.814	0.000
6	MM1	CM3	0.792	0.000
7	HRM3	SCM9	0.765	0.000
8	CM2	CM3	0.758	0.000
9	HRM1	SCM4	0.753	0.000
10	MM4	MM5	0.740	0.000
11	HRM1	HRM3	0.731	0.000
12	HRM7	CM3	0.726	0.000
13	SCM3	MM5	0.726	0.000
14	SCM7	MM4	0.726	0.000
15	HRM7	SCM6	0.722	0.000
16	CM3	CM4	0.717	0.000
17	HRM2	CM3	0.716	0.000
18	SM3	SM4	0.716	0.000
19	HRM2	HRM7	0.713	0.000
20	HRM6	MM2	0.700	0.000
21	HRM6	SM4	0.692	0.000
22	SM1	MM5	0.691	0.000
23	SM3	CM4	0.688	0.000
24	SCM6	CM3	0.674	0.000
25	MM5	CM6	0.669	0.000
26	SCM7	MM5	0.665	0.000
27	SM1	SM4	0.661	0.000
28	SCM3	SCM4	0.658	0.000
29	SM2	SM3	0.657	0.000
30	HRM3	CM1	0.651	0.000

(continued)

S. P. Low and J. Ong, *Project Quality Management*, DOI: 10.1007/978-981-287-074-2,

No.	Factor 1	Factor 2	Pearson correlation coefficient (r)	Sig. (two-tailed)
31	SM3	CM2	0.651	0.000
32	MM3	MM6	0.650	0.000
33	MM4	CM5	0.650	0.000
34	SCM4	SM2	0.648	0.000
35	HRM2	CM2	0.647	0.000
36	HRM7	CM4	0.647	0.000
37	SM1	CM7	0.645	0.000
38	SCM7	MM3	0.643	0.000
39	SM4	MM2	0.641	0.000
40	HRM2	SCM1	0.635	0.000
41	HRM2	CM4	0.630	0.000
42	MM3	MM5	0.629	0.000
43	SCM2	MM4	0.627	0.000
44	SM4	CM3	0.627	0.000
45	MM2	CM3	0.616	0.000
46	MM4	MM6	0.613	0.000
47	CM4	CM5	0.613	0.000
48	SM1	SM3	0.612	0.000
49	HRM1	SM1	0.595	0.000
50	SCM3	CM7	0.594	0.000
51	MM2	CM7	0.593	0.000
52	SM2	CM7	0.584	0.000
53	SM3	CM3	0.584	0.000
54	MM1	MM3	0.579	0.000
55	MM1	CM4	0.579	0.000
56	SCM4	SM1	0.577	0.000
57	HRM2	MM1	0.573	0.000
58	SCM2	MM5	0.573	0.000
59	SCM9	MM3	0.571	0.000
60	CM5	CM7	0.565	0.000
61	HRM7	MM1	0.564	0.000
62	HRM3	HRM6	0.561	0.000
63	MM4	CM3	0.561	0.000
64	SM3	MM2	0.559	0.000
65	SM3	CM7	0.559	0.000
66	MM1	MM2	0.554	0.000
67	HRM6	SM1	0.552	0.000
68	HRM6	SM3	0.552	0.000
69	SCM2	CM5	0.552	0.000
70	SCM2	MM3	0.550	0.000
71	MM3	CM3	0.550	0.000
72	HRM3	SM4	0.549	0.000
73	HRM1	SM2	0.548	0.000
74	MM3	CM5	0.547	0.000
75	SM2	SM4	0.542	0.000

(continued)

No.	Factor 1	Factor 2	Pearson correlation coefficient (r)	Sig. (two-tailed)
76	HRM2	MM2	0.541	0.000
77	SCM9	MM4	0.541	0.000
78	SM2	MM5	0.541	0.000
79	MM2	CM4	0.541	0.000
80	MM6	CM2	0.535	0.001
81	HRM1	HRM6	0.533	0.001
82	SM2	CM1	0.532	0.001
83	SCM5	SCM9	0.531	0.001
84	HRM3	CM6	0.530	0.001
85	SCM1	MM2	0.530	0.001
86	SCM9	MM5	0.525	0.001
87	SM4	MM4	0.525	0.001
88	SM4	CM2	0.525	0.001
89	SCM1	SM4	0.524	0.001
90	MM2	CM2	0.523	0.001
91	CM2	CM4	0.523	0.001
92	SCM5	CM1	0.521	0.001
93	SM1	CM6	0.521	0.001
94	MM1	MM4	0.521	0.001
95	MM3	CM1	0.519	0.001
96	SM1	MM2	0.517	0.001
97	SCM6	SM4	0.516	0.001
98	SM1	CM1	0.514	0.001
99	HRM7	CM2	0.513	0.001
100	SCM8	MM5	0.512	0.001
101	HRM2	SCM6	0.508	0.001
102	SCM3	CM6	0.506	0.001
103	SCM9	CM1	0.504	0.001
104	SM4	CM4	0.498	0.001
105	HRM1	SCM9	0.496	0.002
106	HRM3	SCM6	0.494	0.002
107	HRM1	CM6	0.493	0.002
108	HRM7	MM6	0.492	0.002
109	SM1	MM4	0.492	0.002
110	SM4	MM5	0.492	0.002
111	SCM7	CM1	0.491	0.002
112	MM3	CM6	0.491	0.002
113	HRM3	SCM4	0.487	0.002
114	HRM3	MM2	0.487	0.002
115	SCM4	MM5	0.487	0.002
116	MM2	MM5	0.487	0.002
117	HRM3	SCM3	0.483	0.002
118	SCM9	SM4	0.483	0.002
119	MM4	CM7	0.483	0.002
120	SCM3	SCM8	0.480	0.002

(continued)

No.	Factor 1	Factor 2	Pearson correlation coefficient (r)	Sig. (two-tailed)
121	MM5	CM5	0.478	0.002
122	MM1	CM2	0.477	0.002
123	SCM6	CM4	0.474	0.003
124	SCM3	SM2	0.472	0.003
125	SCM4	SCM9	0.471	0.003
126	MM3	CM7	0.469	0.003
127	CM4	CM7	0.469	0.003
128	MM4	CM1	0.467	0.003
129	SCM7	SCM9	0.465	0.003
130	SCM6	MM1	0.464	0.003
131	SCM6	SCM9	0.463	0.003
132	SCM3	SCM7	0.460	0.004
133	SCM3	SCM9	0.460	0.004
134	SCM5	MM3	0.459	0.004
135	SCM9	CM5	0.458	0.004
136	HRM3	CM7	0.456	0.004
137	HRM1	SCM3	0.455	0.004
138	HRM3	SCM5	0.455	0.004
139	HRM6	SCM1	0.453	0.004
140	SCM3	SM1	0.451	0.004
141	MM2	CM1	0.445	0.005
142	SCM7	MM6	0.444	0.005
143	SCM1	CM3	0.443	0.005
144	HRM1	SCM5	0.441	0.006
145	HRM6	SM2	0.441	0.006
146	HRM2	SM4	0.437	0.006
147	HRM1	CM1	0.435	0.006
148	HRM6	CM7	0.433	0.007
149	SCM8	CM7	0.433	0.007
150	HRM6	CM6	0.432	0.007
151	HRM1	MM2	0.426	0.008
152	HRM3	SM1	0.426	0.008
153	MM6	CM3	0.425	0.008
154	HRM7	SCM1	0.424	0.008
155	SCM2	SCM9	0.422	0.008
156	SCM1	CM2	0.421	0.008
157	SCM3	SM4	0.420	0.009
158	SM4	CM7	0.419	0.009
159	SCM9	CM6	0.418	0.009
160	MM5	CM1	0.418	0.009
161	HRM4	CM1	0.416	0.009
162	HRM6	CM4	0.415	0.010
163	HRM1	CM7	0.414	0.010
164	HRM3	SCM1	0.414	0.010
165	SM4	CM5	0.408	0.011

(continued)

No.	Factor 1	Factor 2	Pearson correlation coefficient (r)	Sig. (two-tailed)
166	MM2	CM6	0.407	0.011
167	SCM8	CM6	0.406	0.011
168	SM3	MM5	0.406	0.011
169	SCM6	SCM7	0.405	0.012
170	SM4	CM1	0.405	0.012
171	SCM3	MM2	0.404	0.012
172	SCM7	CM6	0.404	0.012
173	SM2	CM6	0.404	0.012
174	HRM7	SM3	0.403	0.012
175	MM1	MM5	0.401	0.013
176	CM1	CM7	0.399	0.013
177	SCM7	CM7	0.398	0.013
178	HRM2	SM3	0.394	0.014
179	HRM7	SCM7	0.394	0.014
180	SCM6	SM3	0.393	0.015
181	MM4	CM4	0.390	0.016
182	HRM7	MM2	0.389	0.016
183	SCM9	CM7	0.389	0.016
184	HRM3	CM3	0.387	0.016
185	SCM4	CM6	0.387	0.016
186	MM5	CM3	0.387	0.016
187	CM3	CM7	0.387	0.016
188	MM3	CM2	0.386	0.017
189	MM4	CM6	0.386	0.017
190	SCM3	SM3	0.382	0.018
191	HRM3	SM2	0.378	0.019
192	HRM7	CM1	0.378	0.019
193	SCM1	SCM6	0.378	0.019
194	HRM3	MM5	0.377	0.020
195	SCM3	MM1	0.377	0.020
196	HRM3	HRM7	0.373	0.021
197	CM3	CM6	0.373	0.021
198	SCM2	SM1	0.371	0.022
199	SCM2	CM7	0.369	0.022
200	CM1	CM6	0.368	0.023
201	HRM7	MM4	0.367	0.023
202	CM2	CM6	0.366	0.024
203	SM2	MM2	0.365	0.024
204	SM4	MM3	0.365	0.024
205	SCM7	MM1	0.364	0.025
206	SCM1	MM1	0.363	0.025
207	CM3	CM5	0.363	0.025
208	SCM9	MM1	0.362	0.025
209	MM2	MM4	0.362	0.026
210	SCM4	CM7	0.361	0.026

(continued)

No.	Factor 1	Factor 2	Pearson correlation coefficient (r)	Sig. (two-tailed)
211	SM1	MM3	0.361	0.026
212	HRM7	SM4	0.360	0.027
213	HRM5	HRM7	0.358	0.027
214	SM3	CM5	0.358	0.028
215	SM3	MM6	0.357	0.028
216	HRM2	MM6	0.355	0.029
217	SCM9	SM1	0.351	0.031
218	MM1	CM7	0.350	0.031
219	SCM2	SCM7	0.348	0.032
220	HRM4	MM3	0.346	0.034
221	HRM1	MM5	0.345	0.034
222	SCM7	CM5	0.343	0.035
223	SCM6	MM4	0.342	0.035
224	HRM2	HRM6	0.341	0.036
225	HRM6	CM2	0.341	0.036
226	SCM7	SM1	0.341	0.036
227	SM3	MM1	0.341	0.036
228	SCM7	CM3	0.339	0.038
229	SCM4	SM4	0.338	0.038
230	SCM6	CM2	0.337	0.039
231	SM3	CM6	0.337	0.039
232	HRM3	MM1	0.336	0.039
233	SCM4	SCM6	0.335	0.040
234	HRM7	MM3	0.334	0.040
235	SM1	CM5	0.334	0.041
236	SM4	MM1	0.332	0.042
237	HRM6	CM3	0.330	0.043
238	HRM3	CM4	0.329	0.044
239	HRM1	SM4	0.327	0.045
240	HRM2	SCM2	0.327	0.045
241	MM2	CM5	0.327	0.045
242	SCM7	SCM8	0.324	0.047
243	SCM4	SM3	0.323	0.048
244	MM6	CM1	0.320	0.050

(*Note* Please refer to Table 4.5 for the definitions of the abbreviations in this table)

Appendix F: Statistically Significant Positive Correlation for Usage Rating

No.	Factor 1	Factor 2	Pearson correlation coefficient (r)	Sig. (two-tailed)
1	HRM1	HRM4	1.000	0.000
2	HRM1	HRM7	1.000	0.000
3	HRM1	SCM4	1.000	0.000
4	HRM1	SCM7	1.000	0.000
5	HRM1	SM1	1.000	0.000
6	HRM1	SM2	1.000	0.000
7	HRM1	MM2	1.000	0.000
8	HRM4	HRM7	1.000	0.000
9	HRM4	SCM4	1.000	0.000
10	HRM4	SCM7	1.000	0.000
11	HRM4	SM1	1.000	0.000
12	HRM4	SM2	1.000	0.000
13	HRM4	MM2	1.000	0.000
14	HRM7	SCM4	1.000	0.000
15	HRM7	SCM7	1.000	0.000
16	HRM7	SM1	1.000	0.000
17	HRM7	SM2	1.000	0.000
18	HRM7	MM2	1.000	0.000
19	SCM4	SCM7	1.000	0.000
20	SCM4	SM1	1.000	0.000
21	SCM4	SM2	1.000	0.000
22	SCM4	MM2	1.000	0.000
23	SCM7	SM1	1.000	0.000
24	SCM7	SM2	1.000	0.000
25	SCM7	MM2	1.000	0.000
26	SCM8	MM1	1.000	0.000
27	SM1	SM2	1.000	0.000
28	SM1	MM2	1.000	0.000
29	SM2	MM2	1.000	0.000
30	SCM2	MM5	0.946	0.000

(continued)

S. P. Low and J. Ong, *Project Quality Management*, DOI: 10.1007/978-981-287-074-2,

No.	Factor 1	Factor 2	Pearson correlation coefficient (r)	Sig. (two-tailed)
31	HRM6	CM2	0.851	0.000
32	MM5	CM2	0.851	0.000
33	HRM1	SCM9	0.839	0.000
34	HRM4	SCM9	0.839	0.000
35	HRM7	SCM9	0.839	0.000
36	SCM4	SCM9	0.839	0.000
37	SCM7	SCM9	0.839	0.000
38	SCM9	SM1	0.839	0.000
39	SCM9	SM2	0.839	0.000
40	SCM9	MM2	0.839	0.000
41	HRM2	MM5	0.834	0.000
42	HRM6	SCM2	0.834	0.000
43	SCM6	CM4	0.829	0.000
44	HRM2	CM2	0.805	0.000
45	SCM2	CM2	0.805	0.000
46	HRM6	MM5	0.780	0.000
47	HRM2	SCM2	0.774	0.000
48	HRM2	SCM6	0.774	0.000
49	SCM2	SCM6	0.774	0.000
50	HRM2	MM4	0.764	0.000
51	SCM6	MM4	0.764	0.000
52	SCM3	CM4	0.763	0.000
53	SCM6	MM5	0.723	0.000
54	HRM5	CM6	0.721	0.000
55	MM4	CM4	0.721	0.000
56	SCM3	SCM6	0.715	0.000
57	HRM2	CM4	0.714	0.000
58	MM3	CM1	0.683	0.000
59	HRM1	SCM1	0.678	0.000
60	HRM1	SCM3	0.678	0.000
61	HRM4	SCM1	0.678	0.000
62	HRM4	SCM3	0.678	0.000
63	HRM7	SCM1	0.678	0.000
64	HRM7	SCM3	0.678	0.000
65	SCM1	SCM4	0.678	0.000
66	SCM1	SCM7	0.678	0.000
67	SCM1	SM1	0.678	0.000
68	SCM1	SM2	0.678	0.000
69	SCM1	MM2	0.678	0.000
70	SCM3	SCM4	0.678	0.000
71	SCM3	SCM7	0.678	0.000
72	SCM3	SM1	0.678	0.000
73	SCM3	SM2	0.678	0.000
74	SCM3	MM2	0.678	0.000
75	HRM2	CM1	0.676	0.000

(continued)

No.	Factor 1	Factor 2	Pearson correlation coefficient (r)	Sig. (two-tailed)
76	HRM6	SCM1	0.672	0.000
77	MM5	CM4	0.666	0.000
78	SM4	CM5	0.645	0.000
79	MM5	CM1	0.639	0.000
80	SCM3	MM4	0.638	0.000
81	HRM1	SM4	0.637	0.000
82	HRM4	SM4	0.637	0.000
83	HRM7	SM4	0.637	0.000
84	SCM4	SM4	0.637	0.000
85	SCM7	SM4	0.637	0.000
86	SM1	SM4	0.637	0.000
87	SM2	SM4	0.637	0.000
88	SM4	MM2	0.637	0.000
89	HRM2	CM7	0.622	0.000
90	SCM2	CM7	0.622	0.000
91	SCM6	CM7	0.622	0.000
92	SCM1	SCM3	0.616	0.000
93	HRM2	HRM6	0.611	0.000
94	HRM6	SCM6	0.611	0.000
95	HRM1	CM4	0.600	0.000
96	HRM4	CM4	0.600	0.000
97	HRM7	CM4	0.600	0.000
98	SCM4	CM4	0.600	0.000
99	SCM7	CM4	0.600	0.000
100	SM1	CM4	0.600	0.000
101	SM2	CM4	0.600	0.000
102	MM2	CM4	0.600	0.000
103	SCM2	CM4	0.599	0.000
104	HRM2	SCM3	0.595	0.000
105	SCM1	CM7	0.595	0.000
106	SCM3	CM7	0.595	0.000
107	MM4	MM5	0.592	0.000
108	MM5	CM7	0.588	0.000
109	SCM6	CM2	0.587	0.000
110	SCM9	CM7	0.587	0.000
111	MM6	CM5	0.587	0.000
112	HRM6	CM3	0.582	0.000
113	CM1	CM4	0.580	0.000
114	SCM8	CM6	0.574	0.000
115	MM1	CM6	0.574	0.000
116	CM2	CM3	0.573	0.000
117	HRM1	HRM2	0.567	0.000
118	HRM1	SCM2	0.567	0.000
119	HRM1	SCM6	0.567	0.000
120	HRM2	HRM4	0.567	0.000

(continued)

No.	Factor 1	Factor 2	Pearson correlation coefficient (r)	Sig. (two-tailed)
121	HRM2	HRM7	0.567	0.000
122	HRM2	SCM4	0.567	0.000
123	HRM2	SCM7	0.567	0.000
124	HRM2	SM1	0.567	0.000
125	HRM2	SM2	0.567	0.000
126	HRM2	MM2	0.567	0.000
127	HRM4	SCM2	0.567	0.000
128	HRM4	SCM6	0.567	0.000
129	HRM7	SCM2	0.567	0.000
130	HRM7	SCM6	0.567	0.000
131	SCM2	SCM4	0.567	0.000
132	SCM2	SCM7	0.567	0.000
133	SCM2	SM1	0.567	0.000
134	SCM2	SM2	0.567	0.000
135	SCM2	MM2	0.567	0.000
136	SCM4	SCM6	0.567	0.000
137	SCM6	SCM7	0.567	0.000
138	SCM6	SM1	0.567	0.000
139	SCM6	SM2	0.567	0.000
140	SCM6	MM2	0.567	0.000
141	SCM1	SM4	0.565	0.000
142	SCM1	CM2	0.557	0.000
143	HRM3	CM4	0.552	0.000
144	HRM6	CM4	0.552	0.000
145	SCM2	CM3	0.551	0.000
146	SCM6	CM3	0.551	0.000
147	SCM2	MM4	0.546	0.000
148	CM1	CM2	0.544	0.000
149	HRM2	SCM9	0.542	0.000
150	HRM2	MM3	0.542	0.000
151	SCM2	SCM9	0.542	0.000
152	SCM2	MM3	0.542	0.000
153	SCM2	CM1	0.542	0.000
154	SCM6	SCM9	0.542	0.000
155	SCM6	MM3	0.542	0.000
156	SCM6	CM1	0.542	0.000
157	HRM1	CM7	0.539	0.000
158	HRM4	CM7	0.539	0.000
159	HRM7	CM7	0.539	0.000
160	SCM4	CM7	0.539	0.000
161	SCM7	CM7	0.539	0.000
162	SM1	CM7	0.539	0.000
163	SM2	CM7	0.539	0.000
164	MM2	CM7	0.539	0.000
165	CM2	CM4	0.538	0.000

(continued)

No.	Factor 1	Factor 2	Pearson correlation coefficient (r)	Sig. (two-tailed)
166	HRM1	HRM3	0.536	0.001
167	HRM1	HRM6	0.536	0.001
168	HRM1	MM5	0.536	0.001
169	HRM3	HRM4	0.536	0.001
170	HRM3	HRM7	0.536	0.001
171	HRM3	SCM4	0.536	0.001
172	HRM3	SCM7	0.536	0.001
173	HRM3	SM1	0.536	0.001
174	HRM3	SM2	0.536	0.001
175	HRM3	MM2	0.536	0.001
176	HRM4	HRM6	0.536	0.001
177	HRM4	MM5	0.536	0.001
178	HRM6	HRM7	0.536	0.001
179	HRM6	SCM4	0.536	0.001
180	HRM6	SCM7	0.536	0.001
181	HRM6	SM1	0.536	0.001
182	HRM6	SM2	0.536	0.001
183	HRM6	MM2	0.536	0.001
184	HRM7	MM5	0.536	0.001
185	SCM4	MM5	0.536	0.001
186	SCM7	MM5	0.536	0.001
187	SM1	MM5	0.536	0.001
188	SM2	MM5	0.536	0.001
189	MM2	MM5	0.536	0.001
190	MM4	CM2	0.527	0.001
191	SCM1	SCM9	0.524	0.001
192	SCM3	SCM9	0.524	0.001
193	SCM1	CM4	0.518	0.001
194	MM4	CM1	0.516	0.001
195	CM4	CM7	0.516	0.001
196	HRM6	MM3	0.507	0.001
197	SCM9	MM5	0.507	0.001
198	MM3	MM5	0.507	0.001
199	CM2	CM7	0.501	0.001
200	HRM3	SCM6	0.499	0.001
201	MM4	CM3	0.499	0.001
202	SCM5	SM3	0.487	0.002
203	SCM8	MM5	0.483	0.002
204	MM1	MM5	0.483	0.002
205	SCM9	SM4	0.482	0.002
206	SM3	CM5	0.477	0.002
207	HRM5	CM3	0.476	0.003
208	HRM2	SCM1	0.475	0.003
209	SCM1	SCM2	0.475	0.003
210	SCM1	SCM6	0.475	0.003

(continued)

No.	Factor 1	Factor 2	Pearson correlation coefficient (r)	Sig. (two-tailed)
211	SCM2	SCM3	0.475	0.003
212	MM4	CM7	0.475	0.003
213	SM3	MM6	0.473	0.003
214	MM5	CM3	0.469	0.003
215	SCM3	CM3	0.460	0.004
216	HRM1	CM2	0.456	0.004
217	HRM2	SCM8	0.456	0.004
218	HRM2	MM1	0.456	0.004
219	HRM4	CM2	0.456	0.004
220	HRM7	CM2	0.456	0.004
221	SCM2	SCM8	0.456	0.004
222	SCM2	MM1	0.456	0.004
223	SCM4	CM2	0.456	0.004
224	SCM6	SCM8	0.456	0.004
225	SCM6	MM1	0.456	0.004
226	SCM7	CM2	0.456	0.004
227	SM1	CM2	0.456	0.004
228	SM2	CM2	0.456	0.004
229	MM2	CM2	0.456	0.004
230	HRM6	CM7	0.450	0.005
231	SCM9	CM4	0.444	0.005
232	MM3	CM4	0.444	0.005
233	HRM3	CM5	0.443	0.005
234	SCM3	CM2	0.440	0.006
235	HRM2	CM3	0.436	0.006
236	HRM3	SCM3	0.434	0.006
237	HRM6	SCM3	0.434	0.006
238	SCM1	MM5	0.434	0.006
239	SCM3	MM5	0.434	0.006
240	HRM1	MM4	0.433	0.007
241	HRM4	MM4	0.433	0.007
242	HRM7	MM4	0.433	0.007
243	SCM4	MM4	0.433	0.007
244	SCM7	MM4	0.433	0.007
245	SM1	MM4	0.433	0.007
246	SM2	MM4	0.433	0.007
247	MM2	MM4	0.433	0.007
248	SCM8	CM4	0.431	0.007
249	MM1	CM4	0.431	0.007
250	SCM9	CM2	0.415	0.010
251	MM3	CM2	0.415	0.010
252	MM6	CM6	0.414	0.010
253	HRM1	CM5	0.411	0.010
254	HRM4	CM5	0.411	0.010
255	HRM7	CM5	0.411	0.010

(continued)

No.	Factor 1	Factor 2	Pearson correlation coefficient (r)	Sig. (two-tailed)
256	SCM4	CM5	0.411	0.010
257	SCM7	CM5	0.411	0.010
258	SM1	CM5	0.411	0.010
259	SM2	CM5	0.411	0.010
260	MM2	CM5	0.411	0.010
261	CM3	CM4	0.403	0.012
262	HRM3	SM3	0.394	0.014
263	SCM9	SM3	0.394	0.014
264	SCM9	MM4	0.387	0.016
265	MM3	MM4	0.387	0.016
266	SCM5	CM1	0.382	0.018
267	SCM3	SCM8	0.381	0.018
268	SCM3	MM1	0.381	0.018
269	HRM5	SCM8	0.379	0.019
270	HRM5	MM1	0.379	0.019
271	HRM3	SM4	0.378	0.019
272	HRM6	SM4	0.378	0.019
273	SM4	MM6	0.378	0.019
274	HRM6	MM4	0.377	0.020
275	HRM3	SCM9	0.375	0.020
276	HRM6	SCM9	0.375	0.020
277	HRM6	CM1	0.375	0.020
278	SCM9	CM3	0.372	0.021
279	MM3	CM3	0.372	0.021
280	HRM5	CM5	0.362	0.026
281	CM5	CM6	0.362	0.026
282	SCM8	MM4	0.359	0.027
283	MM1	MM4	0.359	0.027
284	SCM2	SM3	0.357	0.028
285	SCM5	MM6	0.356	0.028
286	SCM2	SCM5	0.355	0.029
287	HRM5	MM4	0.343	0.035
288	MM4	CM6	0.343	0.035
289	CM3	CM7	0.343	0.035
290	SCM1	CM3	0.338	0.038
291	HRM1	SM3	0.331	0.043
292	HRM4	SM3	0.331	0.043
293	HRM7	SM3	0.331	0.043
294	SCM4	SM3	0.331	0.043
295	SCM7	SM3	0.331	0.043
296	SM1	SM3	0.331	0.043
297	SM2	SM3	0.331	0.043
298	SM3	MM2	0.331	0.043
299	SCM8	CM2	0.328	0.045
300	MM1	CM2	0.328	0.045

(continued)

No.	Factor 1	Factor 2	Pearson correlation coefficient (r)	Sig. (two-tailed)
301	HRM5	CM2	0.325	0.046
302	CM2	CM6	0.325	0.046
303	HRM3	MM6	0.323	0.048

(*Note* Please refer to Table 4.5 for the definitions of the abbreviations in this table)

Appendix G-1: Laying of Brickwall Inspection Template

Drawing No: ―――――― Date Inspected: ―――――― Reviewed By: ――――――

Checklist	Conformity (√/X)		Corrective Action Required	Remarks	Who to refer?	Deadline
	S/D	CR				
Preparation & Cleanliness						
Brick Surface						
RC Surface receiving Bricks						
Set Reference Line for Alignment						
Fixing Bonding Bar						
Length of 6mmø M.S. Bonding Bar to Stiffener = 300mm						
3.8mmø Masonry Drive Fixing Pin 28mm Long at 400 c/c						
Spacing between Bonding Bar = 400mm						
Laying of Bricks						
Mixing proportion and use of gauge box						
Use of 64 x 0.5mm thick anodized galvanized steel reinforcement start from 2nd course lay at every 4th course min. lap length of 100mm						
Joint width						
Cavity Wall						
Install 25mm x 3mm thick galvanized steel tie: Start from 4th course lay at every 4th course horizontal spacing 900mm c/c (staggered)						
Provide expansion joint						

S. P. Low and J. Ong, *Project Quality Management*, DOI: 10.1007/978-981-287-074-2,

Alignment & Workmanship						
Surface and edge are plane and vertical						
Crack and chippings						
Pointing						

(S/D: Specifications and Drawings. CR: Constructability Requirements)

Appendix G-2: Prestressed Concrete Structure Inspection Template

Drawing No: ———— Date Inspected: ———— Reviewed By: ————

Checklist	Conformity (√/X)		Corrective Action Required	Remarks	Who to refer?	Deadline
	S/D	CR				
Conditions of Tendons and Anchorages						
Type & Number						
Spacing						
Cleanliness						
Anchorages						
Installation of Sheathing						
Properly Secured						
Free from Damage						
Profile						
Grout Tube						
Stressing and Grouting Operations						
Duct Clean/Free						
Stress/Elongation						
Grouting Operation						
Debonding (For Pre-tensioning)						
Tube Seals at End						
Debond Length						
Materials						

(S/D: Specifications and Drawings. CR: Constructability Requirements)

S. P. Low and J. Ong, *Project Quality Management*, DOI: 10.1007/978-981-287-074-2,

Appendix G-3: Cast In-Situ Concrete Inspection Template

Drawing No: ―――――― Date Inspected: ―――――― Reviewed By: ――――――

Checklist	Conformity (√/X)		Corrective Action Required	Remarks	Who to refer?	Deadline
	S/D	CR				
Formwork						
Dimension						
Level and Plumb						
Tightness of Joints						
Rigidity						
Alignment						
Props & Bracing						
Cleanliness						
Mould Oil						
Reinforcement						
Diameter & Type						
Position & Spacing						
Cleanliness						
Cover						
Hooks & Laps						
Tying Wire						
Rigidity						
Services						
Trimming Bars						
Position & Size						
Cast in Items/Provisions						
Post Concreting						
Curing						
Re-propping						
Formwork Striking						

S. P. Low and J. Ong, *Project Quality Management*, DOI: 10.1007/978-981-287-074-2,

Finished Concrete						
Dimension						
Alignment						
Plumb						
Level						
Honeycomb						

(S/D: Specifications and Drawings. CR: Constructability Requirements)

Appendix G-4: Precast Concrete Structure Inspection Template

Drawing No: ______________ Date Inspected: ____________ Reviewed By: ______________

Checklist	Conformity (√/X)		Corrective Action Required	Remarks	Who to refer?	Deadline
	S/D	CR				
Pre-erection Stage						
Panel Position Lines						
Level & Plumb						
Gridline Reference						
Seal fast Position						
Erection Stage						
Temporary Supports						
Panel Alignment						
Connection Details						
Joint & Seal						
Plumbness						
Cottering Bars						
Connection Welding						
Waterproofing						
Post-erection Stage						
Joint						
Test Results						

(S/D: Specifications and Drawings. CR: Constructability Requirements)

S. P. Low and J. Ong, *Project Quality Management*, DOI: 10.1007/978-981-287-074-2,

Appendix G-5: Structural Steelwork Inspection Template

Drawing No: ———— Date Inspected: ———— Reviewed By: ————

Checklist	Conformity (√/X)		Corrective Action Required	Remarks	Who to refer?	Deadline
	S/D	CR				
Material Check (Visual)						
Element Size						
Element Grade						
Element Condition						
Plumb and Alignment						
Plumb Tolerance						
Line and Level						
Free from Twist						
Connection Plate						
Size of Connection Plate						
Weld Inspection						
Weld Inspection (NDT)						
Bolting						
Size & Type of Bolts						
Required Torque						
Hole Size						
Welding						
Location, Type, Size						
Electrode Type/Size						
Visual Inspection						
NDT						
Welders' Qualifications						
Protection						
Corrosion Protection						
Surface preparation						
Primer						
Painting						
Galvanised						
Anti-skid Coat						
Anti-skid Test						
Temporary Bracing						

(S/D: Specifications and Drawings. CR: Constructability Requirements)

S. P. Low and J. Ong, *Project Quality Management*, DOI: 10.1007/978-981-287-074-2,

Bibliography

Alexander, K., Elke, L., Helmut, K., & Nada, K.-K. (1999). Benchmarking and performance measurement in public sectors: Towards learning for agency effectiveness. *International Journal of Public Sector Management, 12*(2), 121–144.
Allinson, K. (1997). *Getting there by design*. London: Taylor & Francis.
American Concrete Institute. (2005). *The Contractor's guide to quality concrete construction*. St. Louis: American Society of Concrete Contractors.
Arumugam, J. V., & Mojtahedzadeh, R. (2011). Critical success factors of total quality management and their impact on performance of Iranian automotive industry: A theoretical approach. *European Journal of Economics, Finance and Administrative Sciences, 33*, 25–41.
BaniIsmail, L. (2012). An evaluation of the implementation of total quality management (TQM) within the construction sector in the United Kingdom and Jordan. University of Huddersfield.
Bolzano, B., & George, R. (1972). *Theory of science: Attempt at a detailed and in the main novel exposition of logic with constant attention to earlier authors*. Berkeley: University of California Press.
Building and Construction Authority. (2010). CONQUAS Enhancement Series. Retrieved August 20, 2012, from http://www.bca.gov.sg/Publications/EnhancementSeries/enhancement_series.html.
Building and Construction Authority. (2012). Bonus scheme for construction quality. Retrieved August 2, 2012, from http://www.bca.gov.sg/professionals/iquas/conquas_bscheme.html.
Building and Construction Authority. (2012). Precast concrete elements: quality inspection. Retrieved July 20, 2012, from http://www.bca.gov.sg/professionals/iquas/others/precastinspection.pdf.
Chow-Chua, C., Goh, M., & Wan, T. B. (2003). Does ISO 9000 certification improve business performance? *International Journal of Quality & Reliability Management, 20*(8), 936–953.
Concrete Formwork Systems. (1998). Marcel Dekker Incorporated.
Corenet. (2012). Electronic national productivity and quality specifications. Retrieved August 3, 2012, from http://www.corenet.gov.sg/eNPQS/.
Corenet. (2012). Information on construction quality. Retrieved August 6, 2012, from http://www.corenet.gov.sg/iquas/.
Corenet. (2012). List of projects scored under CONQUAS. Retrieved August 14, 2012, from http://www.corenet.gov.sg/homeowners/listsql/default.asp?SortBy=Q29udHJhY3Rvcg==&category=QUxMIENBVEVHT1JJRVM=.
Davis, L. (2002). *Spon's Asia Pacific construction costs handbook*. London: Taylor & Francis.
Dilber, M., Bayyurt, N., Zaim, S., & Tarim, M. (2005). Critical factors of total quality management and its effect on performance in health care industry: a Turkish experience. *Problems and Perspectives in Management, 4*, 220–234.
Druker, J., White, G., Hegewisch, A., & Mayne, L. (1996). Between hard and soft HRM: human resource management in the construction industry. *Construction Management and Economics, 14*(5), 405–416.

S. P. Low and J. Ong, *Project Quality Management*, DOI: 10.1007/978-981-287-074-2,

Dulaimi, M. F., & Hwa, T. F. (2001). Developing world class construction companies in Singapore. *Construction Management and Economics, 19*(6), 591–599.

Husband, S., & Mandal, P. (1999). A conceptual model for quality integrated management in small and medium size enterprises. *International Journal of Quality & Reliability Management, 16*(7), 699–713.

Jarrah, R. T. (2007). Planning for, facilitating, and evaluating design effectiveness. University of Texas at Austin. (The University of Texas at Austin. Civil, A., and Environmental Engineering).

Kline, D. H., & Coleman, G. B. (1992). Four propositions for quality management of design organizations. *Journal of Management in Engineering, 8*(1), 15–26.

Levy, S. M. (2009). Construction process planning and management: An owner's guide to successful projects. Elsevier Science.

Low, S. P., & Abeyegoonasekera, B. (2001). Buildability in design and construction through ISO 9000 quality management systems: Concepts and survey findings. *Architectural Science Review, 44*(4), 355–366.

Low, S. P., & Chia, W. H. (2008). Middle management's influence on the effectiveness of ISO 9000 quality management systems in architectural firms. *Architectural Engineering and Design Management, 4*(3–4), 189–205.

Low, S. P., & Chin, Y. P. (2003). Integrating ISO 9001 and OHSAS 18001 for construction. *Journal of Construction Engineering and Management, 129*(3), 338–347. doi:10.1061/(ASCE)0733-9364(2003) 129:3(338.

Low, S. P., & Peh, K. W. (1996). A framework for implementing TQM in construction. *The TQM Magazine, 8*(5), 39–46.

Low, S. P., & Tan, J. (2005). Integrating ISO 9001 quality management system and ISO 14001 environmental management system for contractors. *Journal of Construction Engineering and Management, 131*(11), 1241–1244. doi:10.1061/(ASCE)0733-9364(2005) 131:11(1241.

Low, S. P., & Tan, W. (1996). Public policies for managing construction quality: The grand strategy of Singapore. *Construction Management & Economics, 14*(4), 295–309. doi:10.1080/014461996373377.

Low, S. P., & Yeap, L. (2001). Quality function deployment in design/build projects. *Journal of Architectural Engineering, 7*(2), 30–39. doi:10.1061/(ASCE)1076-0431(2001) 7:2(30.

Medhat, S. (1997). Concurrent engineering: the agenda for success. Research Studies Press.

Metri, B. A. (2005). TQM critical success factors for construction firms. *Management: Journal of Contemporary Management Issues, 10*(2), 61–72.

Myers, D. (2004). Construction economics: A new approach. Spon.

Neeraj, J. K., & Jha, K. N. (2011). Construction project management. Pearson Education India.

Ofori, G. (1990). The construction industry: Aspects of its economics and management. Singapore University Press.

Quazi, H. A., Hong, C. W., & Meng, C. T. (2002). Impact of ISO 9000 certification on quality management practices: A comparative study. *Total Quality Management, 13*(1), 53–67.

Singh, D., & Tiong, R. (2006). Contractor selection criteria: Investigation of opinions of Singapore construction practitioners. *Journal of Construction Engineering and Management, 132*(9), 998–1008. doi:10.1061/(ASCE)0733-9364(2006) 132:9(998.

Talib, F., & Rahman, Z. (2011). Pareto analysis of total quality management factors critical to success for service industries. Center for Quality.

Tsiotras, G., & Gotzamani, K. (1996). ISO 9000 as an entry key to TQM: The case of Greek industry. *International Journal of Quality & Reliability Management, 13*(4), 64–76.

Turk, A. M. (2006). ISO 9000 in construction: An examination of its application in Turkey. *Building and Environment, 41*(4), 501–511. doi:10.1016/j.buildenv.2005.02.013.

Watson, P., & Howarth, T. (2011). *Construction quality management: Principles and practice*. London: Taylor and Francis.

Yasamis, F., Arditi, D., & Mohammadi, J. (2002). Assessing contractor quality performance. *Construction Management & Economics, 20*(3), 211–223.

MIX
Papier aus verantwortungsvollen Quellen
Paper from responsible sources
FSC® C105338

If you have any concerns about our products,
you can contact us on
ProductSafety@springernature.com

In case Publisher is established outside the EU,
the EU authorized representative is:
Springer Nature Customer Service Center GmbH
Europaplatz 3, 69115 Heidelberg, Germany

Printed by Libri Plureos GmbH
in Hamburg, Germany